SpringerBriefs in Space Development

Series Editor
Joseph N. Pelton Jr.

For further volumes:
http://www.springer.com/series/10058

Virginia E. Wotring

Space Pharmacology

 Springer

Virginia E. Wotring
Johnson Space Center
NASA Rd 1 2101
Houston, TX 77058, USA

ISSN 2191-8171   e-ISSN 2191-818X
ISBN 978-1-4614-3395-8   e-ISBN 978-1-4614-3396-5
DOI 10.1007/978-1-4614-3396-5
Springer New York Dordrecht Heidelberg London

Library of Congress Control Number: 2012932596

© Virginia E. Wotring 2012

This work is subject to copyright. All rights are reserved by the Publisher, whether the whole or part of the material is concerned, specifically the rights of translation, reprinting, reuse of illustrations, recitation, broadcasting, reproduction on microfilms or in any other physical way, and transmission or information storage and retrieval, electronic adaptation, computer software, or by similar or dissimilar methodology now known or hereafter developed. Exempted from this legal reservation are brief excerpts in connection with reviews or scholarly analysis or material supplied specifically for the purpose of being entered and executed on a computer system, for exclusive use by the purchaser of the work. Duplication of this publication or parts thereof is permitted only under the provisions of the Copyright Law of the Publisher's location, in its current version, and permission for use must always be obtained from Springer. Permissions for use may be obtained through RightsLink at the Copyright Clearance Center. Violations are liable to prosecution under the respective Copyright Law.

The use of general descriptive names, registered names, trademarks, service marks, etc. in this publication does not imply, even in the absence of a specific statement, that such names are exempt from the relevant protective laws and regulations and therefore free for general use.

While the advice and information in this book are believed to be true and accurate at the date of publication, neither the authors nor the editors nor the publisher can accept any legal responsibility for any errors or omissions that may be made. The publisher makes no warranty, express or implied, with respect to the material contained herein.

Printed on acid-free paper

Springer is part of Springer Science+Business Media (www.springer.com)

ISU (Society) Page

This Springer book is published in collaboration with the International Space University. At its central campus in Strasbourg, France, and at various locations around the world, the ISU provides graduate-level training to the future leaders of the global space community. The university offers a 2-month Space Studies Program, a 5-week Southern Hemisphere Program, a 1-year Executive MBA and a 1-year Masters program related to space science, space engineering, systems engineering, space policy and law, business and management, and space and society.

These programs give international graduate students and young space professionals the opportunity to learn while solving complex problems in an intercultural environment. Since its founding in 1987, the International Space University has graduated more than 3,000 students from 100 countries, creating an international network of professionals and leaders. ISU faculty and lecturers from around the world have published hundreds of books and articles on space exploration, applications, science and development.

About the Author

Dr. Virginia E. Wotring is a Senior Scientist in the Division of Space Life Sciences at Universities Space Research Association, and Pharmacology Discipline Lead at NASA's Johnson Space Center, Human Health and Countermeasures. She received her doctorate in Pharmacological and Physiological Sciences at Saint Louis University after earning a BS in Chemistry at Florida State University. She has extensive experience studying drug mechanisms of action, drug receptor structure/function relationships, and gene and protein expression.

In 2009, she began directing the NASA JSC Pharmacology Lab, whose mission is to ensure that medications used during spaceflight will act in a predictable, effective and safe manner. The spaceflight environment includes several unusual stressors for the body: microgravity, fluid shifting, increased radiation exposure, lack of circadian cues and others. Each of these can alter normal physiology, and thus, change the way in which drugs are absorbed, distributed, metabolized, excreted or how they interact with physiological targets. Current projects include: determining cardiac safety of new and experimental anti-motion sickness combination therapies, and measurement of gene and protein expression after exposure to spaceflight or a gamma radiation analog of spaceflight.

Dr. Virginia E. Wotring is also actively involved in mentoring students and interns, and participates in teaching at area universities. She holds adjunct appointments in the UTMB Galveston Department of Pharmacology and Toxicology and at the University of Houston's Department of Health and Human Performance.

Contents

1 Introduction.. 1
The Spaceflight Environment... 1
Administration of Medications 3
Basics of Pharmacological Principles 5

2 Absorption 7
Physiological Status.. 10
Spaceflight Evidence... 13
Spaceflight Analog Studies of Effects on Absorption
and Related Factors.. 16
Absorption Summary 17

3 Distribution.. 19
How Local Perfusion Rate Affect Drug Distribution............................. 24
Spaceflight Evidence... 24
Spaceflight Analog Studies of Drug Distribution: Rat Tail
Suspension 25
Spaceflight Analog Studies of Drug Distribution: Bed Rest.................... 25
Distribution Summary.. 26

4 Metabolism and Excretion 27
Spaceflight Evidence... 29
Spaceflight Analog Studies: Rotating Cell Culture 29
Spaceflight Analog Studies: Rat Tail Suspension 30
Excretion... 30
Spaceflight Evidence... 31
Spaceflight Analog Studies: Bed Rest 32
Metabolism and Excretion Summary.. 32

5 Central Nervous System 33
Sleep... 33
    Why Sleep Is Physiologically Required 34
    Sleep Stages 34

Circadian Rhythms... 35
Spaceflight Evidence... 35
What Happens in Sleep Deprivation: Performance
Deficits in General 37
Performance Deficits Specific for Spaceflight 38
Therapies for Sleep Difficulties 38
Risks of Sleep Aid Use Unique to Missions..................................... 39
Potential Countermeasure for Sleep Aid Risks.................................. 40
Central Nervous System Summary 41

6 **Cardiovascular System**.. 43
Spaceflight Evidence... 45
Cardiac Atrophy.. 45
Cardiac Arrhythmias.. 45
Postflight Orthostatic Intolerance 45
Spaceflight Analog Studies.. 47
Cardiovascular Summary 49

7 **Gastrointestinal System**.. 51
Motion Sickness.. 51
Unique Aspects of Space Motion Sickness 52
Nausea and Vomiting 52
The Mechanism of Space Motion Sickness 53
SMS Susceptibility... 55
Spaceflight Evidence... 56
Spaceflight Analog Studies.. 57
Motion Sickness Treatments 58
Problems with Current SMS Therapies 63
Gastrointestinal Summary... 64

8 **Musculoskeletal System**.. 65
Skeletal System.. 65
Spaceflight Evidence... 65
New-Generation Bone-Preserving Drugs: Anti-RANKLs 67
Reduction of Bone Turnover and Reduced Renal Stone Risk 67
Muscular System... 68
Spaceflight Evidence... 68
Testosterone as an Anabolic Steroid 69
New Anabolic Compounds: Selective Androgen Receptor
Modulators 70
Musculoskeletal System Summary 70

9 **Multiple Systems Spaceflight Effects** 71
Immune System 71
Spaceflight Evidence... 71
Do Antibiotics Work Against Microbes Altered by Spaceflight?............ 75
Antibiotic Effect on Native GI Flora 75

Multisystem Radiation Effects 75
    Spaceflight Evidence 75
    Amelioration of Radiation Damage with Pharmaceuticals 76
    Radiation Damage to Stored Pharmaceuticals 80
Multiple Systems Spaceflight Effects Summary 82

10 Conclusions: Special Challenges of Long Duration Exploration 83
Current Drug Testing for Long Duration Spaceflight 83
Unique Medical Requirements for Long Duration Spaceflight 84
Packaging and Shelf Life 84

Abbreviations 87

References 91

Chapter 1
Introduction

One of the risks of space travel that NASA monitors is therapeutic treatment failure. Given that terrestrial medical practices, including therapeutic medications, must be used as the basis for use on missions, there is a possibility the medications will be less than ideal for the actual circumstances encountered on missions. This could be due to either changes in the humans (physiological changes that occur in the spaceflight environment) or changes in the stored medications. This volume is organized into pharmacokinetics and pharmacodynamics, just like many pharmacology textbooks. It addresses each major pharmacokinetic element and each major physiological system in turn, with respect to current and past practices in-flight (Spaceflight Evidence sections) and on the ground (Spaceflight Analog sections).

It is possible that the actions of administered drugs on crewmembers during spaceflight are different from their actions on Earth, but even after more than 40 years of spaceflight experience, most questions about medication use during missions remain unanswered. Use of medications with insufficient knowledge about their actual results may result in inadequate treatment and may even reduce human performance and well-being in some circumstances. There is evidence that this has already occurred during and immediately after spaceflights. The spaceflight pharmaceutical results knowledge base must be improved to enable flight surgeons and crewmembers to make better-informed decisions about using pharmaceuticals during flight.

The Spaceflight Environment

The spaceflight environment induces changes in human physiology, and these changes have been the subject of much study over the past few decades. However, these studies are confounded by the small number of subjects, as well by the inability to separate the different stressors of spaceflight (radiation exposure and microgravity, for example) from each other. In every physiological system, the details of spaceflight-induced physiological changes are not completely understood.

V.E. Wotring, *Space Pharmacology*, SpringerBriefs in Space Development, DOI 10.1007/978-1-4614-3396-5_1, © Virginia E. Wotring 2012

Despite this fact, crewmembers are treated with pharmaceuticals to reduce or prevent medical problems, with little information about drug function in their altered physiological systems.

There are two major concerns about pharmaceutical use in the unusual environment of spaceflight. The first is that the actions of pharmaceuticals on physiology altered by a spaceflight environment are currently assumed to be the same as the actions in terrestrial use. This assumption is based on anecdotal reports of acceptable treatment outcomes and has not been tested in a systematic fashion. The wide range of physiological systems altered by spaceflight and the degree of change experienced in some of them make it seem likely that alterations in pharmaceutical action will be seen. As the duration of missions lengthens to include more distant exploration, it becomes more likely that problems will be encountered. The second concern is that the integrity of stored pharmaceuticals must be established to ensure that adequate amounts of active compounds are available in each dose and that degradation to toxic compounds is minimized. This risk also depends on mission duration, as longer exploration-class missions will probably not include opportunities for supply replenishment and will require that drugs be stored much longer than their usual terrestrial shelf lives.

In spaceflight, the body undergoes a broad spectrum of changes, and historically, crewmembers have complained of a number of ailments. Space motion sickness, sleep disturbances, sinus congestion, headache, body pain, and minor infections have been the most commonly reported health problems (Putcha et al. 1999; Clement 2003).

Pharmacological interventions are currently used to alleviate the immediate symptoms that impair crew function, as well as to minimize long-term damage to crewmembers' bodies (Putcha et al. 1999). Drugs and dosages have been determined empirically, with terrestrial medical practices as a guide, particularly practices from military and commercial flight medicine. It has been assumed that drugs will act on the body in spaceflight the same way they would act on the body on Earth. This hypothesis has not been properly tested for all of the drugs currently used in space, or for additional drugs that may be required for longer journeys.

The human body has evolved to function well on Earth, exposed to a gravitational field of 1 G and protected from most cosmic rays by Earth's atmosphere. When the body is removed from gravitational forces and the atmosphere, many changes are seen (Table 1.1). Gastrointestinal motility is reduced. The otoliths of the inner ear no longer rest on hair cells, and fail to send meaningful signals to the brain regarding motion and proprioception. Bone is broken down, with increased urinary calcium resulting in formation of renal stones. Muscle, especially underused muscle, atrophies. Cardiac abnormalities such as arrhythmias may occur. Exposure to radiation increases the risk of DNA damage that could lead to cell death or cancer. Immune system function may diminish. These changes have been described, but currently none is fully understood. Many questions about the mechanisms that initiate and control these changes remain unanswered.

Furthermore, gravity is not the only force at work. Motion-induced sickness is common in terrestrial travel and certainly plays a role in space travel. Astronauts are in a high-stress job, and flights are the most critical times for their performance to

Table 1.1 A 1991 list of possible factors contributing to uncertainties in using pharmaceutical treatment. The role (or lack thereof) of each factor is still largely unknown (Santy and Bungo 1991) Used with permission SAGE Publishing

Factors Affecting Pharmacological Efficacy in Space	
Operational	
	Circadian shifting of crews
	Incorrect or inadequate dosages
	Incorrect dosing regimens
	Use of multiple medications
	Cabin life support alterations
Physical	
	Physical adaptation to the space flight environment
	SMS symptoms
	GI motility changes
	Altered nutritional or energy requirements
Physiologic	
	Physiologic adaptation to the spaceflight environment
	Physiologic changes in normal drug metabolism
	Fluid compartment changes
	Alterations in indices of physiologic stress

be at peak. The confinement of the launch and landing situations makes bathroom trips impossible, so astronauts tend to voluntarily (and perhaps unconsciously) reduce intake on those days, perhaps contributing to fluid loss. After the acute stresses of launch are over, other factors become more important. Radiation exposure may contribute to compromised immune function. Concurrently, less than ideal sanitation leads to increased exposure to one's own microbes as well as those from fellow crewmembers. Lack of privacy and separation from family both lead to emotional stress, which may have physical ramifications. Lack of circadian cues, as well as lack of a private, quiet, and dark sleep area, contribute to difficulty sleeping. This can further compromise physical well-being. All of these additional factors surely play their roles in affecting space travelers.

Some of the issues thought of as being attributable to low gravity may actually be caused by the elevated gravity conditions astronauts must pass through on the way to low gravity, specifically the high-g periods of launch and landing. This may explain why some microgravity models (also called spaceflight analogs) don't seem to effectively reproduce the environment that crewmembers actually experience.

Administration of Medications

In typical clinical practice on Earth, patients consult with their physicians and obtain advice on what medications they should use and how these medications should be used. Medication is then dispensed by a pharmacist, enough to treat the patient according to the physician's advice. Records of the drug and recommended dosage are kept by both medical professionals. It is then assumed that the patient will follow the usage directions. Crewmembers are in an unusual situation in that they have been provided with a pre-packed field kit of medications meant to meet the needs of several people for several months during spaceflight. They have the opportunity to consult with ground-based flight surgeons, but they also have free access to the medication kit and can self-administer as they see fit. This means that records of medication use may not always be complete. Attempts are made to capture information about medication use with postflight debriefings, but it is understandably difficult for crewmembers to recall all medication use, indications, and side effects extending back over weeks or months.

The recent data-mining activity of the JSC Pharmacotherapeutics group provides an example of the type of data currently available for researchers. It has been assumed that in-flight drug efficacy differs from ground-use efficacy, but this question has never been directly addressed in a research study. In an attempt to use the information that existed in crewmembers' medical records to determine the in-flight efficacy of space motion sickness (SMS) treatments, the JSC Pharmacotherapeutics group requested that all available information on the topic be collected by the Medical Informatics group and de-identified for research use. The resulting database included portions of medical debriefs from 511 crewmembers on 88 shuttle missions from STS-1 through STS-94.

The first important finding was that only 62% of the crewmembers responded to their medical debrief questionnaires. Furthermore, of these reports, at least 35% were incomplete. The magnitude of the missing data makes it very difficult to draw conclusions from any findings. However, the data from 387 total doses of SMS medications taken by 132 crewmembers were analyzed. Each reporting crewmember took an average of 1.9 doses per flight; the high per-person was 9 doses. Fifty-five of these doses were taken prophylactically on flight day 0 (launch day). Promethazine, with or without dexedrine, was the most common choice, and was generally reported to be effective (Fig. 1.1) (Putcha 2009). However, given that information from about half the crewmembers was unavailable, it is difficult to give this result much weight. It simply is not known if the other crewmembers were not troubled by SMS, took medications with satisfactory results, used medications that did not work well for them, frequently repeated doses, combined SMS drugs with other kinds of drugs, and so on.

The aim of this document is to evaluate the current scientific literature regarding pharmaceutical use in space, and to determine which areas are in greatest need of further research to reduce the risks associated with pharmaceutical use in spaceflight. The role of the JSC Pharmacology Discipline is to ensure that crewmembers and their physicians have the best possible information regarding the likely action of pharmaceuticals during spaceflight. The selection of drugs, including choices of preparations and strengths, as well as usage guidelines and dosing, all fall under the purview of the clinicians in Flight Medicine and the Clinical Pharmacy.

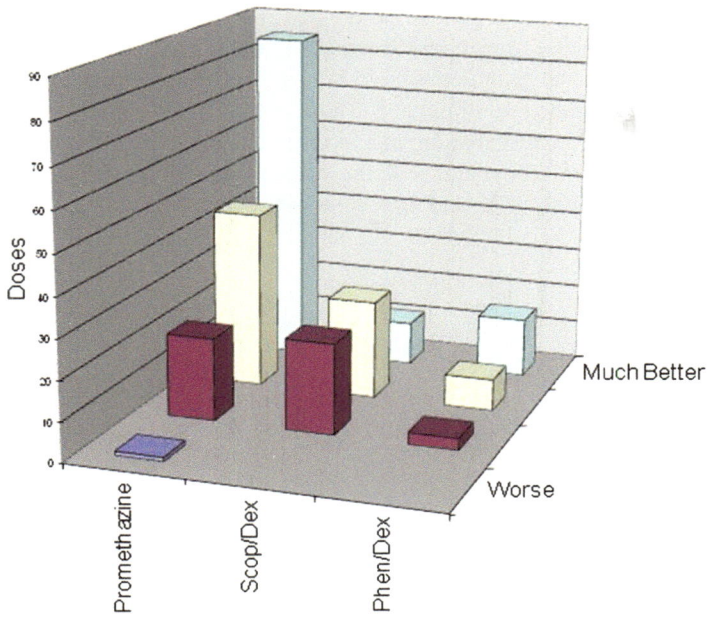

Fig. 1.1 Comparison of space motion sickness drug use on Shuttle flights. The y-axis indicates the number of doses used of each drug listed on the x-axis. Self-reported efficacy is shown on the z-axis (Putcha 2009). NASA, Open Access

Basics of Pharmacological Principles

Pharmacology is the study of drugs and how they act on the body. Any chemical that can interact with the body can be considered a drug or pharmacological agent, although the tendency is to include only those agents administered intentionally in order to achieve some beneficial therapeutic effect, such as treating an illness or promoting healing of an injury. (Chemicals taken in unintentionally or those that produce undesirable effects are usually thought of as toxins, and their study is called toxicology (Gilman et al. 1990)). This definition of "drug" includes food, drinks, supplements, and herbs in addition to over-the-counter medications, prescription drugs, and recreational drugs.

Medications, pharmacological agents, and therapeutic agents are interchangeable terms for those drugs used to cure disease or alleviate symptoms. All medications are drugs, but some drugs are not used therapeutically and are not considered medications. For example, nicotine and alcohol are currently considered drugs but not medications, although in the past, each has been used therapeutically and would once have been considered a medication (Gilman et al. 1990).

The study of pharmacology can be divided into two major subfields, pharmacokinetics and pharmacodynamics. Pharmacokinetics is the study of how the body acts on the drug as the drug is absorbed into the body, transported throughout the body, and then degraded or eliminated. These concepts are known by the acronym ADME, or absorption, distribution, metabolism, and excretion. The other aspect of pharmacology, pharmacodynamics, is the study of how drugs act on the body and interact with its tissues, cells, and molecules to produce the desired effects.

Chapter 2
Absorption

Absorption is the process by which a drug leaves its site of administration and gets delivered into the bloodstream. The speed and extent of absorption depend on the route of administration, and on the physical and chemical properties of the drug. These include its formulation, which involves not only whether the drug is a pill, capsule, liquid, or aerosol, but also the binders, coatings, and flavors used in a particular preparation (Gilman et al. 1990). The physiological status of the body also plays a significant role in absorption. The state of hydration, whether the stomach is full or empty, and the fat content of the last meal can each directly affect absorption of drugs from the gut, and may also indirectly influence gastrointestinal motility. Spaceflight alters many aspects of human physiology, directly and indirectly. Thus it has the potential to significantly affect absorption, and this must be carefully weighed when considering giving medications in flight.

A variety of administration routes are possible, including oral (PO), intravenous (IV), intramuscular (IM), subcutaneous (SC), transdermal (TD), rectal (PR), and intranasal (IN). The administration route is chosen by the physician and depends on properties of the drug, the particular formulations available, the susceptibility of the drug to fast inactivation by liver enzymes (that is, first-pass metabolism) as well as the patient's abilities and preferences (Table 2.1).

Bioavailability is a concept closely associated with absorption that refers to the amount of the drug that is eventually available to the target tissues. It is largely determined by the formulation of the drug: what the active ingredient may be mixed with, packaged in, or coated with to make a particular pill, cream, or other dosage form. These formulation details are mostly determined by inactive ingredients, but they are important elements affecting absorption of the active ingredients and are carefully considered during drug design and testing. Different formulations with the same dose may deliver very different bioavailabilities (Kopacek 2007). For example, for a relatively hydrophobic compound, an oral dose may require a pill of 800 mg to achieve dissolution in the aqueous gastrointestinal fluids and subsequent delivery of 400 mg to the circulation. For the same drug, it may be possible to use a transdermal patch of 400 mg to allow delivery of 400 mg to the circulation.

V.E. Wotring, *Space Pharmacology*, SpringerBriefs in Space Development, DOI 10.1007/978-1-4614-3396-5_2, © Virginia E. Wotring 2012

Table 2.1 Routes of drug administration

Routes	Advantages	Disadvantages
IV	Fastest effect	Must be administered by trained personnel
	No first-pass metabolism	Risk of injury, esp in flight
	100% bioavailable	For water soluble drugs only
IM	Fast effect	Can be self-administered with training
	70–100% bioavailable	
IN	Convenient	Few drugs available in this form
	5–100% bioavailable	Slow effect
PO	Convenient	Slow
	5–100% bioavailable	First-pass metabolism may be significant
TD	Convenient	Very slow onset
	80–100% bioavailable	Few drugs available in this form
	Prolonged absorption	

Along the same lines, a tablet coated to be less irritating to the stomach may have to be made at a higher dose to arrive at the same plasma concentration as an uncoated tablet.

The route of administration used most frequently is oral (PO). Some agents can be given by an intranasal (IN) or transdermal (TD) route conveniently and with good result. Intramuscular (IM) administration works well for many agents, but requires training of the patient. For good effect other agents must be given intravenously (IV), which essentially bypasses the absorption issue, making this route excellent for emergency situations and for drugs with very poor water solubility. However, this method requires a well-trained practitioner and steady surroundings, and also carries a risk of injury (Gilman et al. 1990; Kopacek 2007). In flight, PO and IM have been the most frequently used routes of administration (Putcha 1999). Since it is feasible to teach crewmembers to give themselves IM injections, these injections are more practical during spaceflight than in the typical patient population.

Chemical properties of a drug affect its absorption. Water-soluble drugs dissolve quickly and are distributed among body tissues as freely as water. In general, the more hydrophobic a compound, the more slowly it will be dissolved and distributed. The degree of water solubility of a drug is also influenced by pH; weak acids are absorbed while in their non-ionized form in the stomach. Thus pH is a strong determinant of absorption location: some agents are more soluble in the acidic conditions of the stomach, whereas others remain insoluble until the pH increases after they leave the stomach. Hydrophobic drugs are more bioavailable if delivered transdermally or in a slow IV. Although TD administration provides slow, continuous delivery of drug, it may be too slow if the condition requires fast onset of action. The IV route is routinely used for emergency situations. Some hydrophobic drugs are orally delivered; dosage is generally increased to allow for incomplete absorption. Hydrophilic drugs are much more versatile and can be formulated for delivery via several routes, with oral being the most typical because it is most acceptable to patients (Gilman et al. 1990; Kopacek 2007).

The actual site of absorption depends on both the administration route and the chemical properties of the drug. Except in the case of IV administration, the drug will need to cross several semipermeable membranes, each of which may have its own qualities. Tight junctions between cells have different permissiveness in different tissues, and some membranes contain embedded proteins designed for transport of small molecules. Additional barriers are found in some tissues, such as the mucosal layer of the gastrointestinal (GI) tract and the dead cell layer of the skin. Drug molecules may cross membranes by a variety of mechanisms. Passive absorption may occur, especially with small uncharged molecules. Facilitated or active transport proteins in a membrane may recognize specific chemical interaction sites on a drug molecule (that are similar to the sites for which these transporters were designed) and carry drugs across membranes. Proteins are often carried into cells by a pinocytotic mechanism along with water (Gilman et al. 1990; Kopacek 2007).

Orally administered drugs that are lipid soluble or weak acids are readily absorbed in the stomach. However, not many drugs are absorbed in the stomach because of its low pH and small surface area, and the relatively short contact time. Protein drugs are almost immediately denatured by the acidity of the stomach. The mucosal layer of the stomach is relatively thick, which physically limits exposure of the contents to the membrane. However, in general, drugs are absorbed mostly in the small intestine, because of its large surface area and the relatively long time they are exposed to it. Along the length of the small intestine, the pH varies from about 4 at the duodenum to 8 at the lower ileum, meaning that somewhere along that continuum the pH is likely to be favorable for any ionizable compound (Gilman et al. 1990; Kopacek 2007). For research purposes, oral acetaminophen is the recognized standard for absorption, since it is readily and quickly absorbed by passive diffusion in the small intestine (Clements et al. 1978).

There are tremendous differences in rate and site of medication absorption among the different oral dosage forms. Enteric coatings render tablets less soluble in the acidic conditions of the stomach, which is desirable if the medication is irritating to the stomach tissue or sensitive to acidic degradation. Extended-release dosage forms use a variety of coatings and excipients to prolong release rates. This is beneficial for maintenance of constant circulating concentrations for chronic treatment, but can reduce or delay plasma concentration after the administration of the initial dose (Shargel et al. 2005).

The physical and chemical properties of the administered pharmaceutical and the physiological state of the body are not the only factors that determine the ultimate plasma concentration of active ingredient. In cases where patients administer the drug on their own, patient compliance, or adherence to instructions, can play a significant role. Patient compliance failures usually result in under-administration of drugs. On average, patients self-administer less than 80–85% of their total prescribed dose (Kass et al. 1987; Cramer et al. 1989; Kruse et al. 1991; Kruse et al. 1992; Kruse et al. 1993; Saini et al. 2009). In some cases, physicians will intentionally overprescribe to compensate, usually by indicating a higher dose than actually required (Urquhart and Vrijens 2006). Less frequent dosing (sometimes with higher doses) is now thought to be acceptable in more cases than previously thought, since for certain drugs, the mechanism of drug action involves initiation of a cascade of events that continues for some time period after the drug concentration falls below therapeutic levels (Mattie et al. 1989). It has been shown that there is a great improvement in compliance with three or fewer doses per day (1/day, 87%; 2/day, 81%; 3/day, 77%) compared to 4 per day (39%) (Cramer et al. 1989). However, compliance rates are only a little better with a single dose per day than with 2 or 3 doses per day. A number of monitoring systems are under development to measure and/or improve compliance, most using electronic tags on medication bottles (Kruse and Weber 1990; Kruse et al. 1991, 1992, 1993; Vrijens and Urquhart 2005; Takacs and Hanak 2008), but no such system has been implemented for use on missions. Currently, the only way to know what crewmembers are taking during flight is to ask them, usually in the form of a medical consultation or a research study.

All of these factors that influence plasma concentration of administered drug may make it seem that achieving constant therapeutic concentrations is next to impossible. Most clinically used drugs can also be toxic in the wrong circumstances; in fact it is often said that any drug can become a poison at the wrong dose. This notion is taken into account during drug development and approval. The drugs used as medicines are generally effective and safe over some range of plasma concentrations; this is why the same dosing is typically used even for patients with large differences in body mass. Most FDA-approved drugs have a relatively wide margin of safety, which pharmacologists call "therapeutic index" and define as the LD_{50}/ED_{50}. This is the ratio of the dose at which 50% of patients would be killed (LD_{50}, extrapolated from animal experiments) and the dose at which 50% of patients experience the desired therapeutic effect (ED_{50}). The slope of the concentration-response curve is a rough indicator of therapeutic index; drugs with higher slopes have narrower therapeutic indices. Drugs with a therapeutic index less than 3 are considered narrow, and prescribing physicians are advised to closely monitor adverse effects and even circulating plasma concentrations in their patients. The safest drugs are the ones with no overlap between the therapeutic curve and the toxic curve, typically a therapeutic index of least 10.

The fact that medications can be both safe and effective over a range of circulating concentrations leads to flexibility in dosing and dose-timing. This fact may mean that crewmembers who may suffer weight loss over the course of a mission and chronic or transient fluid loss at various points during a mission may not need to alter medication dosing (particularly for drugs with a wide therapeutic index), although this is another point that has yet to be directly tested.

Physiological Status

Several physiological factors can influence absorption. Hydration state will affect absorption of drugs administered by any route, whereas GI motility, blood flow, pH, microfloral environment, and vomiting will particularly affect orally administered drugs. These effects are comprehensively reviewed by Fleisher and colleagues (Fleisher et al. 1999).

Hydration state affects the plasma concentration of a drug in that if the plasma volume is reduced, there is an apparent increase in the plasma concentration. This effect tends to more strongly influence the elimination phase of the concentration-time curve, as the drug concentration decreases over time. It has been well demonstrated in veterinary medicine that in animals dehydrated such that their body weight is reduced by 7–12%, peak concentrations of administered drugs are virtually unaffected, but elimination half-lives are doubled to quadrupled, respectively (Elsheikh et al. 1998). Few of these experiments have been conducted in human subjects; this literature is largely derived from North African domestic animals, which are probably well adapted to dry conditions. Regardless, the magnitude of chronic dehydration typically seen in spaceflight is about 1–2% of body weight, with only transient increases at launch.

Gut motility ultimately determines the residence time of gut contents, which directly affects absorption. Motility is under nervous system control, largely through muscarinic acetylcholine receptors (Crema et al. 1970). Scopolamine is an anti-nausea drug that blocks these receptors (Katzung 2007), reducing motility. Although it is not known if motility is directly affected by microgravity, motility certainly would have been affected by the space motion sickness (SMS) treatments (scopolamine and promethazine) that have been used during flights (Wood et al. 1987; Wood et al. 1990; Davis et al. 1993a). Records of scopolamine dose timing and amount are not available, and thus it is difficult to interpret its effectiveness as a motion sickness treatment and any associated untoward effects seen in previous flight studies.

Ground-based evidence would predict that after an initial dose of scopolamine, gastric emptying would be slowed, which would result in slower progression to the intestine of any medications and nutrients taken subsequently. Weakly acidic drugs may show increased absorption as their time in the stomach is increased, but for most medications, particularly basic ones, absorption occurs more readily in the more central regions of the intestine. Anti-muscarinic treatment will delay absorption, or even reduce it if stomach acid has a degradative effect. This altered absorption capability could last for many hours depending on the dose given.

The presence of food in the gut will affect absorption of orally administered drugs, through several mechanisms. Food affects the rate of gastric emptying, with differential effects for different food types. Food itself may interact with a drug, changing its activity or absorption (McLachlan and Ramzan 2006; Smith et al. 2009a). An excellent example of this is the chelation of tetracycline by calcium, resulting in the patient instruction about avoiding calcium-rich foods with tetracycline doses (Kakemi et al. 1968a, b). The presence of food in the stomach triggers release of gastric acid and digestive enzymes and increases local blood flow, all part of the digestive process and factors that can affect drug absorption (Dressman et al. 1993). Taking a drug with food (within 2 h) will slow the time to peak concentration and will decrease the maximum drug concentration. However, the area under the curve, which is a measure of the total amount of a drug present over time, is not changed by food in the gut, so this finding does not usually affect patient instructions (Oosterhuis and Jonkman 1993). The presence of food in the gut also causes increased gastrointestinal (GI) and liver blood flow, which can alter the metabolism

Fig. 2.1 Mean serum amoxycillin levels after the administration of amoxycillin (250 mg) to normal male volunteers. Note that bed rest and sleeping (*both supine positions*) show nearly identical results, very different from what was seen in the vertical ambulatory group. From Roberts and Denton (1980) Used with permission, Springer

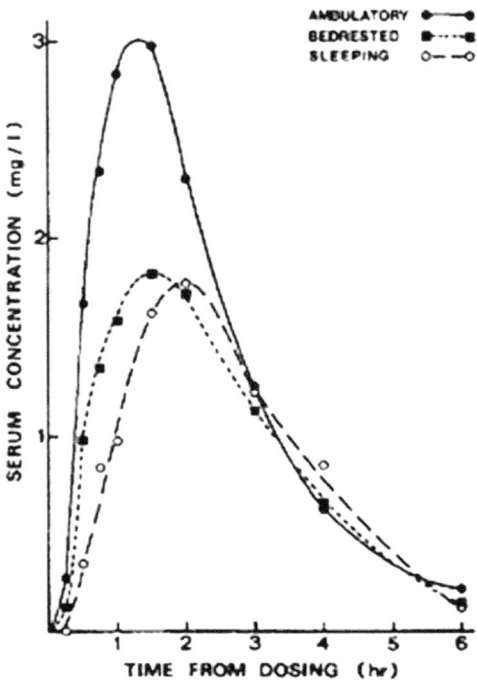

of drugs that undergo hepatic metabolism. This will be discussed in more detail in the metabolism section below.

The effect of body position on gastric function is a well-studied phenomenon, first reported in 1918 (Debuys and Henrique 1918). It is well known that body position or posture affects the rate of gastric emptying (Rumble et al. 1991; Oosterhuis and Jonkman 1993; Queckenberg et al. 2009). In a study comparing ambulatory subjects with those lying down (either sleeping or during a single day of bed rest) changing body position was shown to affect the peak plasma concentration of a drug (Fig. 2.1) (Roberts and Denton 1980). In a different study, the time to peak plasma concentration after dosing with acetaminophen was slowed by 50% by lying on the right side compared to lying on the left side or standing, likely because positioning of the stomach contents over the pylorus increased gastric emptying in the latter two conditions (Renwick et al. 1992). The direct effect of gravity, or lack thereof, on gastric function has not yet been well studied, but from the results discussed above, it follows that gravity must play some role, if not the major role, in the effects reported to be caused by body position. Before the initial human spaceflights of the 1960s, there was speculation as to whether food consumption and digestion would be possible in microgravity, but this was quickly dismissed when John Glenn consumed the first food on orbit (Smith et al. 2009b).

Evidence exists that the predominant species of the gut microflora shift during spaceflight, that is, some species may decrease in number while others increase (Shilov et al. 1971). In general, it is clear that microflora can affect drug absorption (Fleisher et al. 1999; Schneeman 2002); however, it has not yet been determined whether the changes that have been seen in flight significantly affect absorption.

Spaceflight Evidence

Few published studies exist regarding the effects of spaceflight on drug absorption, and these are covered in the excellent review by Gandia and colleagues (Gandia et al. 2005). Limited flight data are available, from case studies of a very few individual astronauts (Cintron et al. 1987a; Cintron et al. 1987b; Putcha et al. 1996). Much more thorough studies have been performed in head-down-tilt bed rest, but the validity of this model compared to flight has not been determined with respect to absorption (Gandia et al. 2003).

Acetaminophen is an acknowledged standard measure of pharmacokinetic (PK) absorption (Clements et al. 1978), and its typical time to peak is about 1 h after oral administration, with half-life of about 2 h (Clements et al. 1978; Ameer et al. 1983). Cintron et al. (1987a) reported two astronaut case studies that show lower and slower peak concentrations of acetaminophen early in flight (Fig. 2.2). The number of replicates is small for any human study, and frequency of sampling is low for this type of kinetic study (both of these parameters are affected by the mission). Also, it is not clear if the subjects were free of other drugs. Data from three individuals (A, B, C) show tremendous variability (Fig. 2.2). Individuals B and C had a reduced peak concentration on flight day 1–2. Individual A showed much slower absorption in the baseline preflight trial. Unfortunately the paucity of data and variability in data preclude drawing a definitive conclusion, but note the similarity of these results to the slower rate and decrease in peak seen in supine subjects (Fig. 2.1) (Roberts and Denton 1980).

A similar reduction in the kinetics of absorption was shown in a more comprehensive recent study. Acetaminophen tablets were administered to five "long-duration crewmembers" (mission day not specified) on the International Space Station (ISS) who were free of other drugs. A 60% slower time to peak concentration was seen (Fig. 2.3). The area under the curve (AUC), a measure of total drug absorbed, was unchanged. When the acetaminophen was administered as capsules rather than tablets, the time to peak increased by 30%, again with no change in AUC (Kovachevich et al. 2009). The changes in time to peak might reflect absorbance changes due to position of the pill or capsule in the stomach. On Earth, a capsule has a small amount of air trapped inside and tends to float on top of the liquid contents of the stomach, whereas tablets sink toward the pylorus. Away from the influence of Earth's gravity, either type of medication would be expected to behave differently in the stomach. However, because the total amount of medication absorbed did not change, the only difference being in the timing of the peak concentration, alterations in prescribing are probably not warranted.

Fig. 2.2 Salivary
acetaminophen before and
during spaceflight in three
different individuals (Cintron
et al. 1987a). NASA, Open
Access

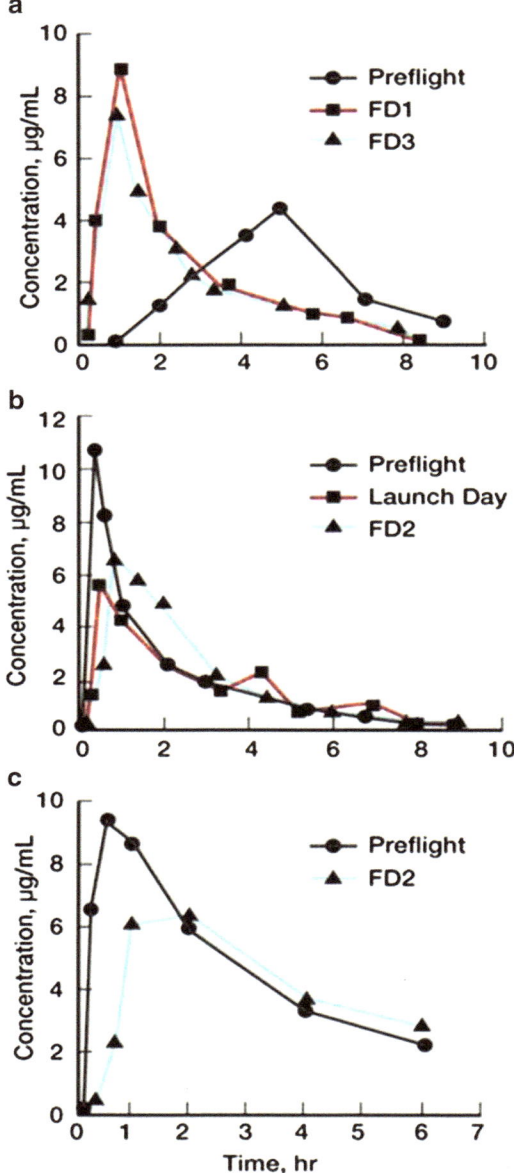

Cintron et al. have published additional case studies from flight using scopolamine collected from saliva, again from three individuals and using less than optimal sampling frequency (Fig. 2.4) (Cintron et al. 1987b). Scopolamine is notorious for erratic oral absorption and is not a good tool for evaluating the absorption process (Pavy-Le Traon et al. 1994; Saivin et al. 1997), but at the time of this experiment, it was of considerable clinical interest. With this caveat in mind, the peak concentration results were indeed highly variable, but it is interesting to note that in every case,

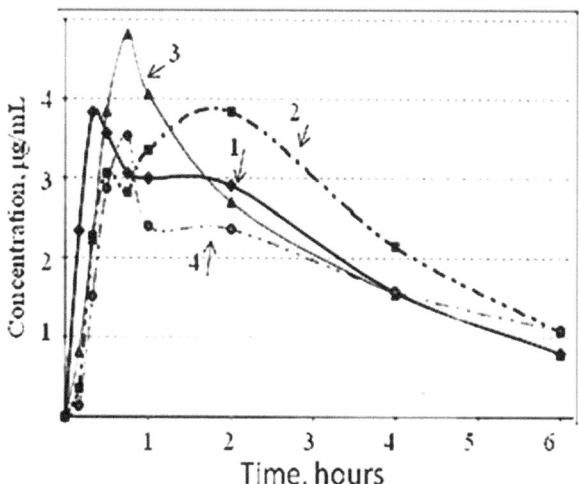

Fig. 2.3 Average pharmacokinetic curves of acetaminophen. 1, tablets, on ground; 2, tablets during spaceflight; 3, capsules, ground; 4, capsules during spaceflight. Note that the spaceflight effect differs for capsules and tablets (Kovachevich et al. 2009). Used with permission, Springer

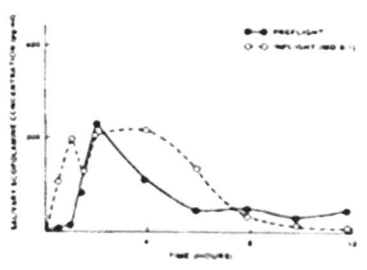

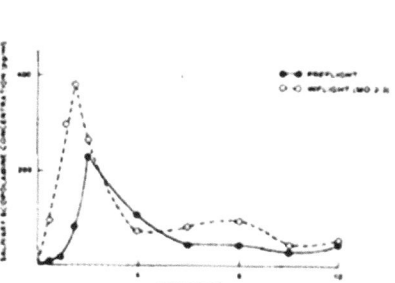

Figure 3. Salivary concentration-time profiles of scopolamine following oral administration of 0.4 mg scopolamine and 5 mg dextro amphetamine to a crewmember.

Cintron & Putcha, 1987

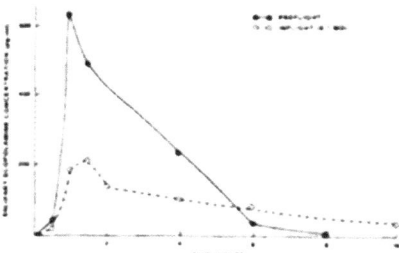

Figure 5. Salivary concentration-time profiles of scopolamine following oral administration of 0.4 mg scopolamine and 5 mg dextro amphetamine to a crewmember.

Figure 4. Salivary concentration-time profiles of scopolamine following oral administration of 0.4 mg scopolamine and 5 mg dextro amphetamine to a crewmember.

Fig. 2.4 Salivary scopolamine concentrations before and during spaceflight. The two *upper panels* are from the same crewmember; the *lower panels* are two additional individuals, measured on the flight day shown in the legend (Cintron et al. 1987b). NASA, Open Access

peak drug concentration was reached after about 2 h, whether scopolamine was administered before, during, or after flight (Fig. 2.4) (Cintron et al. 1987b). In another low gravity model, parabolic flight, a study showed that scopolamine in a gelcap formula reached peak plasma concentration about 1 h after administration, 15–20 min slower than a tablet preparation (Boyd et al. 2007), but it should be noted that this model includes periods of elevated gravity as well as microgravity, and their effects cannot be separated.

These few in-flight case studies seem to indicate that oral absorption is no less variable in space than it is on Earth. It is clear that for a conclusion to be drawn, additional data would have to be collected. However, these preliminary results (mostly in the form of case studies) do not indicate that drug absorption is dramatically affected by spaceflight.

Spaceflight Analog Studies of Effects on Absorption and Related Factors

Bed rest for days to months, typically at a head-down tilt of 6°, has been used extensively as a model of weightlessness for over half a century. It mimics (albeit imperfectly) the fluid shifts that crewmembers experience in flight, which is the chief attraction for its use in drug absorption studies (Pavy-Le Traon et al. 2007). Subjects in bed rest are generally inactive, particularly with regard to the large muscles used to support the body in a standing position and maintain balance, another way in which the model mimics spaceflight (Krasnoff and Painter 1999). However, the effects of bed rest or inactivity on gastrointestinal (GI) function have not been well studied in healthy individuals, so this model should not be regarded as thoroughly validated for pharmacokinetic studies. Furthermore, the fluid shift and dehydration observed in bed rest studies do not replicate flight data (Drummer et al. 1993; Norsk et al. 1995; Leach et al. 1996; Johansen et al. 1997).

At least partial success has been achieved with attempts to improve bed rest as an analog for spaceflight by supplementation of thyroid hormone, but this updated model has not been widely adopted (Ito et al. 2010; Lovejoy et al. 1999). Stress and metabolic effects also seem to be inconsistent between the analog and actual spaceflight, which may confound comparisons of drug metabolism in bed rest and flight. As a result, bed rest data must be critically evaluated with these issues firmly in mind.

An 80-day bed rest study was conducted by Gandia and colleagues in 2001 (10 subjects) and 2002 (8 subjects) (Gandia et al. 2003). Subjects were given 1 g acetaminophen, in capsule form, on bed rest days 0 (pre-bed rest), 1, 18, and 80. Samples of plasma and saliva were collected over time for analysis of drug concentration. They showed that plasma concentration peaked earlier as time in bed rest increased; the time to peak was shortened from about 2 h to less than 1 h. Similarly, the peak concentration increased with time in bed rest, approximately doubling at the 80-day administration. The saliva results were virtually identical to those from plasma (Gandia et al. 2003). This study is in direct conflict with a study of supine (for 2 h before dosing) compared to ambulatory subjects ($n = 8$) in which a 10- to 15-fold slower time to peak acetaminophen concentration was found (Nimmo and Prescott 1978). Whether the difference can be explained by dosage form or degree of tilt (supine or −6° head-down) is not known.

A study with six volunteers showed that absorption of ciprofloxacin was not significantly affected by 3 days of bed rest (Schuck et al. 2005). Unfortunately, 3 days of bed rest was the only time point measured in this study, so these data may be of limited utility, although this drug is an important antibiotic.

Absorption Summary

In a worst-case scenario, if absorption increased dramatically, particularly in the case of a drug with a narrow therapeutic index, peak plasma concentration could be elevated into the toxic range, resulting in overdose. Alternatively, if low GI blood flow or vomiting resulted in no drug absorption, it would be as if the drug was never administered and the original complaint would remain essentially untreated. These scenarios are unlikely, but could be life-threatening or mission-compromising.

Furthermore, a significant number of anecdotal reports strongly indicate no significant changes in drug absorption in spaceflight. Acetaminophen is the gold standard for absorption measurements, and it has been given therapeutically (rather than scientifically) many times during flight. There have been no reports from crewmembers or flight surgeons that acetaminophen was ineffective, or that it caused overdose symptoms.

It is not known if absorption of drugs is altered in flight. The available spaceflight evidence is observation-based, as opposed to hypothesis-driven (Category III). Controlled studies have contributed evidence from spaceflight analogs including bed rest (Category II), but the validity of these models with respect to drug absorption has not been established. A systematic record of crew symptoms, therapy used, perceived effectiveness, and description of side effects could be extremely useful for deciding if further absorption studies are required for particular drugs.

Chapter 3
Distribution

Distribution is the process by which a drug leaves the bloodstream and is delivered to body tissues, including the target organ. Diffusion is typically uneven throughout the body, chiefly because of differences in perfusion rate in the various organs and tissues. Once absorbed, most drugs do not spread evenly throughout the body. Hydrophilic drugs tend to stay within the blood and the interstitial space. Hydrophobic drugs tend to concentrate in fatty tissues. Other drugs may concentrate mainly in only one small part of the body if the tissues there have a special attraction for and ability to retain the drug. Regional differences in pH and membrane permeability also play a role in differential distribution. However, in theory, a drug will distribute from the main circulation into capillaries and then equilibrate into extracellular spaces and then intracellular spaces of organs and tissues. This movement of a drug throughout the body occurs through body water compartments after a drug is dissolved and absorbed.

Total body water measures 42 L in a 70-kg person, or about 60% of the total body weight. The largest single component of total body water (28 L, 40% of body weight) is the fluid inside the cells of the body, intracellular fluid. Each cell has multiple mechanisms by which it controls intracellular fluid volume, osmolarity, and concentrations of individual ions and compounds such as adenosine triphosphate (ATP) and glucose, in order to maintain homeostasis. About one-fourth of the total volume (11 L) is in the spaces between cells, called interstitial fluid. The plasma (the cell-free portion of blood) in circulation has a volume of about 3 L of water.

In addition to plasma, blood contains water that is inside the cells of the blood, for a total of 5 L of water in the blood. All of these compartments (intracellular, interstitial, and plasma) are physically separated by cellular membranes and are osmotically connected, meaning that water will cross the membrane to equalize osmolarity between neighboring spaces, and small molecules (like many drugs) may be able to cross as well. Other, smaller water compartments in the body are more separate from the other body water because the specialized membranes defining them do not permit much movement of water or other molecules. These transcellular spaces total 1–2 L and include the ocular fluid, the synovial fluid in joints, the pericardial fluid, and the cerebrospinal fluid.

V.E. Wotring, *Space Pharmacology*, SpringerBriefs in Space Development, DOI 10.1007/978-1-4614-3396-5_3, © Virginia E. Wotring 2012

Some of these fluid spaces can be measured relatively directly: total body water by antipyrine or labeled water, extracellular fluid by inulin, plasma volume by Evans blue dye, radio-labeled tracers, or carbon monoxide rebreathing. Volumes of the other spaces must be derived mathematically from those that can be measured. Blood volume is considered an important clinical factor in cardiovascular functions and is usually calculated from plasma volume measurements (Guyton and Hall 2006). These volumes have been measured before and after spaceflight.

Body water is constantly in flux, with fluid continuously being consumed and excreted. The drive to consume water is regulated as well as the amount of water being excreted as urine or retained by the kidney. Regulatory and sensing mechanisms exist in several physiological systems that coordinate general homeostasis with the needs and conditions of specific organs and tissues. To regulate systemic osmolarity, hypothalamic osmoreceptors increase pituitary secretion of antidiuretic hormone (ADH) into the circulation when they sense an increase in osmolarity. ADH signals the kidney to reabsorb more water and produce less urine volume. This increases the volume of the extracellular space, which is sensed by baroreceptors in the atria and the carotid artery, in the upper part of the body. When a person lies down, extracellular body fluid shifts upward, which is interpreted by these baroreceptors as an increase in extracellular fluid, and they respond via the renin-angiotensin-aldosterone system, signaling the kidney to excrete additional water (Berne and Levy 1988).

It was long assumed that microgravity would also initiate this regulatory cascade, leading to diuresis and subsequent dehydration, which would then cause a suite of symptoms including headache, impaired GI motility, and orthostatic hypotension (Gauer and Henry 1963), but limited in-flight measurements have not supported this (Leach et al. 1996; Norsk et al. 2000). It has been proposed that the baroreceptor-kidney loop does not participate in fluid volume regulation in microgravity the way it does on Earth (Hargens and Richardson 2009). The baroreceptor-kidney loop of fluid volume regulation in Earth gravity involves mainly the vascular part of the extracellular space, but fluid volume regulation in microgravity is proposed to be driven by a decrease in gravity-driven hydrostatic pressures that shift extracellular fluid from blood vessels to the interstitial space. However, the data upon which this argument was built may be missing the critical time windows near launch and landing. Operational demands mean that crewmembers are unavailable for fluid measurements or urine collection during the periods when the greatest changes would be expected. This constraint has led to extrapolations from data collected several days before and several days after launch. Consequently, this is a controversial area.

It has been well established that gravity affects tissue fluid distribution on Earth: it causes blood to pool in the lower limbs, exerting a greater hydrostatic force on capillary walls there and resulting in a shift of fluid from the vasculature to the interstitial space (Fig. 3.1) (Hargens and Watenpaugh 1996). This hydrostatic force would be absent in microgravity (Diedrich et al. 2007); thus, the normal gravity-driven hydrostatic pressures on the capillary walls in the lower body would be decreased. This would permit more equal distribution of vascular fluid around the body, which would be perceived as an upward fluid shift. In the first few hours

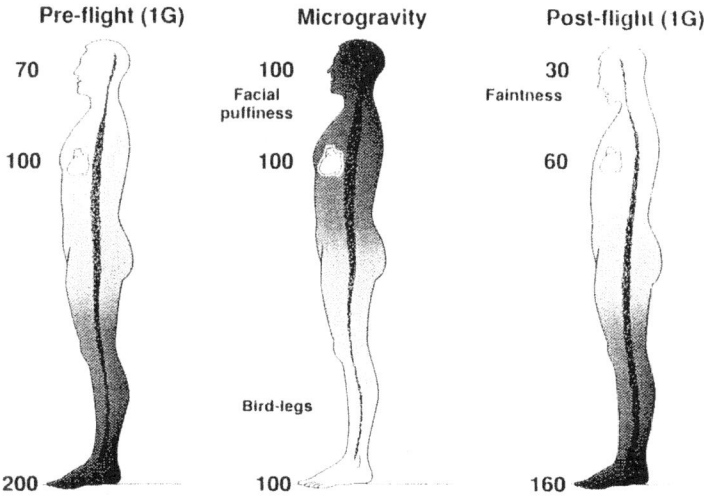

Fig. 3.1 Hypothesized mean arterial pressures due to gravity and loss of gravity before, during, and after microgravity exposure. Note the apparent increase in pressure at the head during flight, and loss of pressure at the head and torso after landing (Hargens and Watenpaugh 1996). Used with permission, Wolters Kluwer Health

of weightlessness, the volume of the lower limbs does decrease while thoracic volumes increase (Montgomery et al. 1993). Blood volume decreases, but the mechanism for this is unclear. Over a period of a few days, the body seems to adapt to this new condition; crewmembers typically do not report long-term fluid problems (Hargens and Watenpaugh 1996). One study suggests that the volume of intracellular fluid increases (Leach et al. 1996).

Currently, there is no convincing evidence of chronic fluid losses larger than 15% during spaceflight. Although a 15% fluid loss may affect the function of the cardiovascular system, no apparent effects on absorption or distribution of administered drugs have been reported. The larger, transient fluid shifts in the first few days of flight cause acute symptoms of head congestion and swelling, an increase in intraocular pressure, and other effects, most of which seem to resolve in a matter of hours to a few days.

The question of how similar these transient fluid shifts of spaceflight are to fluid shifts that occur in bed rest complicates the use of bed rest as a flight analog for pharmacology experiments. In a recent review, Diedrich and colleagues (Diedrich et al. 2007) compared the plasma volume effects of the supine position on Earth to those of spaceflight (Fig. 3.2). In the supine position on Earth, the hydrostatic pressure caused by blood pooling in the legs is reduced, central (thoracic) blood volume increases, and the baroreceptor-kidney loop of the fluid regulation system causes diuresis. The authors propose that in space, fluid shifts are minor and transient, diuresis is actually reduced (Drummer et al. 1993), and total fluid volume remains relatively constant. It should be noted that the data supporting these statements were extrapolated from pre- and postflight measurements.

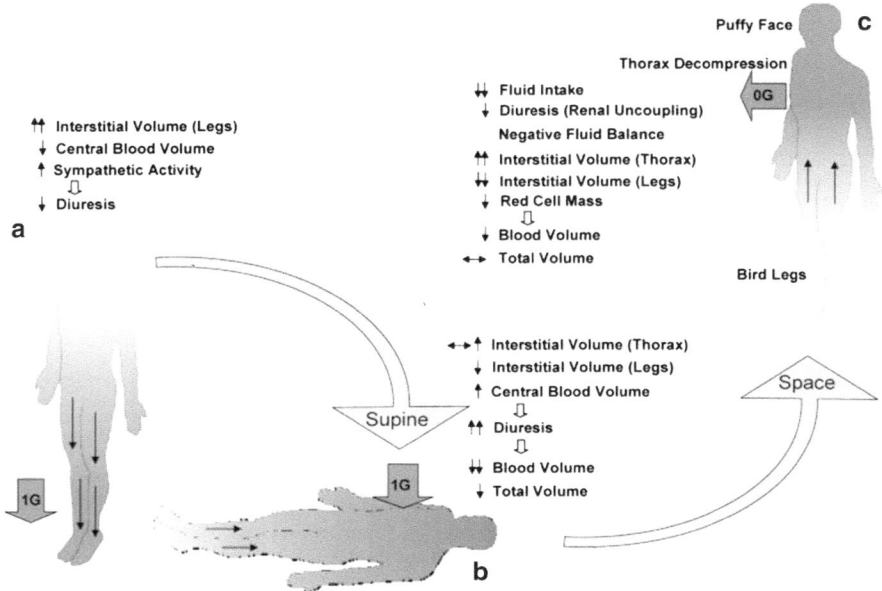

Fig. 3.2 Diagrammatic representation of the volume regulation in 1G upright (**a**), 1G supine (**b**), and microgravity (**c**), from Diedrich et al. (2007). Used with permission, Elsevier

More recent reviews of bed rest studies (Meck et al. 2009; Platts et al. 2009a) are in disagreement with these conclusions, although the data are similar. A transient diuresis at the initiation of bed rest was seen, with a 15% reduction in plasma volume maintained until return to ambulation (Meck et al. 2009; Platts et al. 2009a). Measurements of arterial function in dependent and independent vessels showed decreases in intimal medial thickness (maintained over the duration of bed rest) as well as an increase in hyperemic reactivity (Platts et al. 2009a). These findings support the notion that blood flow is altered during bed rest. Some studies have attempted to optimize the bed rest scenario to make it more comparable to spaceflight. Most notably, the administration of thyroid hormone seems to create a negative nitrogen balance more like that seen in spaceflight (Lovejoy et al. 1999).

There are a wealth of conflicting reports on the subject of fluid shifting in space-flight. The subject is confounded by use of the term "fluid shifting" to refer to fluid loss or diuresis and also to the redistribution of fluid toward the head and away from the lower extremities. The subject is also beset by a preponderance of descriptive reports as opposed to studies designed to test hypotheses and possible mechanisms. Also, it seems that the brief hypergravity experienced at liftoff and landing may also play a role in fluid shifting (Sumanasekera et al. 2007), which makes comparisons of flight data and flight analog data problematic, since the flight analogs are simulations of microgravity alone.

Existing evidence suggests that the initial, brief hypergravity stress of liftoff starts a cascade of events beginning with a transient increase in permeability of cell membranes, particularly those of vascular endothelium (Sumanasekera et al. 2007). This would permit fluid and possibly small molecules to cross these membranes more freely than they ordinarily would. Red cell mass then decreases, possibly by neocytolysis (Rice et al. 2001). Diuresis would decrease, in contrast to the increase seen in the bed rest model (Drummer et al. 1993; Diedrich et al. 2007; Meck et al. 2009; Platts et al. 2009a). Estrogens have also been shown to affect the permeability of tight junctions (Cho et al. 1998); it is possible that chronic permeability increases occur that are related to environmental estrogens or other hormonal changes in flight.

It may seem counterintuitive that fluid shifts or apparent losses do not play a significant role in effective concentration of administered drugs. The safety factors (such as the therapeutic index) built into drug development and approval seem to accommodate all but the most extreme fluid shifts, but because, in terrestrial medicine, the most extreme fluid shifts are immediately rectified by fluid administration, the actual extent of the effects of any fluid loss on pharmacokinetic parameters is not well established in humans. Few publications exist on the specific topic of pharmacological treatment of the chronically dehydrated patient (with dehydration as a sole medical problem). Studies have been conducted on treatment of the elderly, who sometimes suffer from chronic dehydration in addition to reduction in hepatic and renal function, as well as other concurrent conditions (Turnheim 2003; Turnheim 2004), but because of the number and variety of the other conditions, these data must be considered carefully before extrapolation to healthy adults who have suffered a spaceflight-induced loss in body fluids.

Terrestrially, the generally accepted treatment plan for anyone displaying symptoms of dehydration is first to treat the dehydration by fluid administration, then to use typical dosing strategies for any other indicated drugs, as opposed to accommodating the dehydration by altering dosing. For these reasons, there is little evidence in the literature regarding drug distribution in the chronically dehydrated person. There is, however, an extensive veterinary literature on the treatment of North African domestic animals, although it must be considered with care because these animals, having lived in arid conditions for many generations, are likely to be extremely well-adapted to hot and dry conditions, unlike most humans. In camels dehydrated by 10% of their body weight, the AUC and maximum concentration of febantel, an antihelminthic, were lower (Ben-Zvi et al. 1996). Other studies have been conducted in more typical laboratory animals. Water deprivation (24–48 h) in Lewis rats has been shown to have no effect on the serum and cerebrospinal concentrations of IV-administered theophylline, although there were increases in serum proteins and hematocrit (Zhi and Levy 1989). Changes in the LD_{50} of dextroamphetamine have been reported with dehydration, although neither the degree of dehydration nor the methods used to cause it were mentioned in this report (Muller and Vernikos-Danellis 1968). It is difficult to extrapolate from studies like these to the astronaut population. Additional studies with crewmembers may be required to know if the degree and duration of fluid shifting experienced by crewmembers during missions requires alterations in medication administration.

How Local Perfusion Rate Affect Drug Distribution

Local perfusion rate may be altered by fluid shifting, because there is less fluid in the vasculature and more in tissues. Tissue fluid is not necessarily moving; it is not pumped by the heart, but is pushed around by the action of surrounding muscles. However, in microgravity, in the absence of an exercise program, the large standing balance muscles perform less action, and thus less force pushes on tissue fluid. Local blood flow has been shown to be decreased during venous stasis, likely by local sympathetic reflexes (Henriksen and Sejrsen 1976). Local fluid dynamics can also play a role in absorption (GI circulation) and clearance of drugs (hepatic and renal circulation) (Rowland 1975). In addition to these, the direct impacts of this finding on drug distribution remain untested.

Spaceflight Evidence

Experiments designed to test the distribution of drugs in body fluids or tissues in microgravity have not been reported. However, many pre- and postflight measurements have been made of a variety of cardiovascular variables, including volumes of some of the body water compartments. It should be noted that mission constraints prohibit these kinds of measurements for at least 24 h before launch or landing, and a similar period after; it remains possible that significant shifts (losses, gains, or compartment changes) could occur during these prohibited times and not be measured.

A few body water experiments have been done during flight. In four crewmembers on a 5-day shuttle flight, total body water (measured by isotope dilution using ^{18}O) decreased by about 3% in spaceflight (Leach et al. 1991b). The same research group expanded the study on a later flight and found that extracellular fluid volume and plasma volume dropped by about 10% in flight, especially in the first 2 days of flight, while intracellular fluid volume increased by 10% and total serum protein was constant (Leach et al. 1996).

Sodium excretion dropped, but water was not retained as it would be on Earth (Leach et al. 1996). Furthermore, most crewmembers use a variety of medications, mostly for prevention or treatment of SMS symptoms, during the launch and landing time windows; no mention is made in the Leach study of subjects refraining from medication use. Another study has noted that both vasopressin and renin are increased in spaceflight (Drummer et al. 2000b). Taken together, these results provide support for a model that includes a fluid shift on flight day 1–2, upward in the body and from the plasma, interstitial, and extracellular spaces into the intracellular spaces. There is no convincing evidence regarding the distribution of drugs during or after any fluid shifts.

Spaceflight Analog Studies of Drug Distribution: Rat Tail Suspension

The tail-suspended rat model is used by physiologists as a cardiac deconditioning model and by those interested in spaceflight as means to mimic fluid redistribution of flight in a small laboratory animal. Tail-suspended rats showed no meaningful changes in tissue distribution of IV-administered ^{3}H-nicotine compared to ambulatory controls (Chowdhury et al. 1999). Nicotine passes easily through plasma membranes and tends to distribute quickly throughout the body tissues.

Perhaps more directly relevant to drug distribution, tail-suspended rats show a transient increase in serum protein, including albumin, on day 1, followed by a decrease on day 3 (Brunner et al. 1995). This finding means that protein binding may be transiently increased, which can reduce drug availability for that time period. Promethazine is routinely taken during this time period, and normally it is up to 93% bound by circulating protein. A boost in plasma protein could reduce the effective promethazine dose, but this notion has not been directly tested.

Spaceflight Analog Studies of Drug Distribution: Bed Rest

Because basic physiological variables are routinely monitored in most bed rest studies, whether head-down tilt (HDT) or supine bed rest is used, much information is available about fluid volumes in bed rest. Total body water was decreased in bed rest (Czarnik and Vernikos 1999). In supine bed rest, women lost 10–15% of their lower-body volume, whereas men lost 0–7% (Montgomery 1993). The hormones involved in systemic water regulation have been shown to be affected by this analog. Renin, vasopressin, and aldosterone all increased considerably in HDT bed rest (Maillet et al. 1995). These studies are the first steps toward understanding the mechanisms involved in fluid redistribution in bed rest, and possibly in spaceflight if this model can be reliably validated.

Pharmacokinetic studies are more feasible in bed rest than in spaceflight, but the drugs that have been studied in this model have thus far been drugs of clinical interest rather than the best tools for evaluation of pharmacokinetic parameters. In a study comparing an oral dose of amoxicillin during sleep, bed rest, and ambulation, increased renal clearance was shown, without a change in absorption. Lying down, either for a long time in bed rest or during sleep, decreased the peak plasma concentration and slightly decreased the time to peak (Roberts and Denton 1980). A study of plasma and tissue concentrations of orally administered ciprofloxacin showed that plasma concentrations were identical before and after 3 days of HDT bed rest. Tissue concentrations were not significantly different, either (Schuck et al. 2005). Both amoxicillin and ciprofloxacin are of clinical but not pharmacokinetic interest. Unfortunately, the most pertinent study for evaluation of drug distribution was conducted in supine bed rest instead of HDT. Plasma concentrations of IV-administered lidocaine and penicillin were unchanged by supine bed rest for 9–10 days. The same investigators showed that protein binding with lidocaine showed no significant changes after bed rest (Kates et al. 1980). This is a compelling result since lidocaine is considered the pharmacokinetic gold standard for measurements of drug distribution, especially for determining the extent of protein binding (Saivin et al. 1997). However, the degree to which drug distribution in a supine body position mimics the drug distribution in spaceflight is unknown.

Distribution Summary

No distribution evidence from flight studies is available; therefore, no ground model for distribution can be validated. Therapeutic use of drugs during flight has been reported to be largely effective (Putcha et al. 1999), suggesting that most drugs must be distributed in a near-normal fashion (Category III, observational evidence). However, because of concerns about fluid shifting and the potential for dehydration, it would be prudent to determine if drug distribution is altered during flight. Flight experiments with classical distribution probes (erythromycin, propranolol (Saivin et al. 1997)) would likely provide the data needed to establish or rule out drug distribution as a potential flight problem. A systematic record of crew symptoms, therapy used, perceived effectiveness, and description of side effects could be extremely useful for deciding if additional distribution studies are required for particular drugs.

Chapter 4
Metabolism and Excretion

Metabolism refers to the biochemical reactions required to maintain life, but in general it is typically used to refer to those reactions leading to the energy-harvesting breakdown of ingested food. The effect of spaceflight on food metabolism has been the subject of much research (Yegorov et al. 1972; Rambaut et al. 1977a, b; Abraham et al. 1980; Macho et al. 1982; Macho et al. 1991a; Yamaguchi et al. 1991; Jiang et al. 1993; Tischler et al. 1993; Macho et al. 1996; Wade et al. 2000; Macho et al. 2001; Wade et al. 2002; Macho et al. 2003; Zwart et al. 2004; Zwart et al. 2005; Smith and Zwart 2008). Many studies have shown that during spaceflight humans lose weight, particularly in the form of bone density and muscle mass (Meehan 1971; Adey 1972; Yegorov et al. 1972; Leach et al. 1983; Smith et al. 1999; Drummer et al. 2000b; LeBlanc et al. 2000; Drummer et al. 2001; Zwart et al. 2004; Cavanagh et al. 2005; LeBlanc et al. 2007; Zwart et al. 2009). Alterations in the hormones that control salt and water balance, cell growth, and immune function have been found in spaceflight (Leach et al. 1972; Leach 1979; Leach 1981; Leach et al. 1983; Leach et al. 1988; Cintron et al. 1990; Leach 1991; Leach et al. 1991; Drummer et al. 2001; Wade et al. 2002; Santucci et al. 2009). In animals, altered insulin levels and changes in glucose tolerance have been seen (Bernardini and Taub 1969; Macho et al. 1991b; Song et al. 2002; Tobin et al. 2002). Various countermeasures aimed at improving reduction-oxidation balance or indicators of physiological stress have been tested in animals (Chowdhury et. al. 2007). It is clear that spaceflight has multisystem effects on the body.

In pharmacology, metabolism refers specifically to biochemical reactions that break down administered pharmaceuticals or alter pharmaceuticals so they can be eliminated by the body (Gilman et al. 1990). The ability of humans to metabolize and clear drugs is a natural process that involves the same enzymatic pathways and transport systems that are utilized for food metabolism and homeostasis. Most pharmaceuticals are hydrophobic compounds that pass easily through lipid membranes. This results in rapid absorption by diffusion, but creates a problem for the body as it attempts to clear xenobiotics from the system. If the hydrophobic compounds can be partitioned to the urine for excretion, they are simply absorbed across membranes back into the circulation instead of being excreted in the urine. For the body to rid itself of these chemicals, they must be made more hydrophilic, so that they will tend to remain in the urine until excretion. This is typically accomplished by the addition or unmasking of a hydrophilic group, which can occur in the intestinal wall or in the liver. Phase I reactions include reduction-oxidation and hydrolysis reactions, and most involve enzymes of the cytochrome P450 family.

Some drugs remain biologically active after Phase I reactions, but are subsequently inactivated when conjugated to endogenous substances in Phase II reactions such as sulfation or glucuronidation. Hundreds of enzymes, mostly in the liver, are involved in transforming xenobiotics for elimination, and their activities may be affected by the physiological state of the body (Leucuta and Vlase 2006). Metabolism results in the inactivation of drugs' therapeutic effectiveness and facilitates elimination. (The opposite is true for those drugs that are administered in a pro-drug inactive form, whose activation requires metabolism.)

The extent of metabolism can determine the efficacy and toxicity of a drug by controlling its biological half-life. Among the most serious considerations in the clinical use of drugs are adverse drug reactions. If a drug is metabolized too quickly, it rapidly loses its therapeutic efficacy. If a drug is metabolized too slowly, the drug can accumulate in the bloodstream; as a consequence, the pharmacokinetic parameter AUC (area under the plasma concentration–time curve) is elevated and the plasma clearance of the drug is decreased. These enzymes also show tremendous genetic variability, giving rise to significant differences in metabolic rates among individuals (Gilman et al. 1990). The effect of spaceflight on the activity of specific enzymes has been investigated, but information in this area is by no means complete.

Polypharmacy is a problem of growing concern in terrestrial medicine, since more of the population are using multiple medications. Some drugs are known to alter the rate of metabolism and clearance, which can result in what is effectively overdosing, or in delivery of insufficient drug to the target organ (Katzung 2007). For example, it has been shown that antifungal azole derivatives interact with cytochrome P450 enzymes in a fashion that decreases clearance of zolpidem, a sleep aid used on space missions (Greenblatt et al. 1998a). On Earth, the potential for polypharmacy interactions is monitored by the prescribing physician and the clinical pharmacist, but in current mission operations, opportunities for this sort of monitoring are made more difficult.

Drug metabolism is further complicated by natural genetic variation. It is now known that different individuals may express various forms of many of the key metabolic enzymes, and some isoforms have faster activities than others. This has led to the new field of pharmacogenomics (Lee et al.), which is now being used to tailor drug dosing regimens for specific individuals based on the DNA sequences that encode their metabolic enzymes (sometimes called "personalized medicine" in the popular press). Currently, this type of variation has been noted for only a single drug used in spaceflight (codeine), but as more genetic polymorphisms are identified and new drugs are added to the formulary, it may become prudent to test crewmembers to determine metabolic reaction rates and prepare optimized treatment plans for individual crewmembers according to their genotypes.

Spaceflight Evidence

In rats that experienced spaceflight for 14 days, morphological analysis showed that hepatocytes were larger than those of vivarium or synchronous controls, although the livers themselves were not larger (Racine and Cormier 1992). In rats flown on Spacelab 3 (7 days), a decrease of about 50% was seen in total P450 enzyme activity (Phase I metabolism), whereas no change occurred in glutathione S-transferase, a Phase II enzyme (Hargrove and Jones 1985). In rats, after an 8-day flight on STS-63, a reduction was seen in the amount of the liver enzymes catalase and glutathione reductase (both involved in general antioxidant activity), as well as GSH sulfur-transferase, a Phase II enzyme (Hollander et al. 1998). It is not known if enzyme concentration or amount correlates well with enzyme activity for these enzymes in these conditions.

It is also difficult to know how the various flights may have differed. One research group reports no changes in rat P450 enzymes after one 14-day flight (Merrill et al. 1992), 15% reduction after a different 14-day flight (Merrill et al. 1990), and ~50% reduction after a 7-day flight (Merrill et al. 1987). Measurements of radiation exposure on each of these flights would be very helpful, but these data were not reported. In the 1992 paper, a delay of several hours in retrieving the animals after landing is mentioned, and the authors suggest that measured changes would have been larger if there had been no delay (Merrill et al. 1990).

Although at first glance it would seem that the spaceflight environment would be the same for animals and humans on the same spacecraft, this may not be the case. Details of the animals' housing arrangements are not mentioned in every study, but both Macho (Macho et al. 1982) and Allebban (Allebban et al. 1996) describe a 12-h light period followed by 12 h of darkness, and Merrill et al. say that their animals experienced 16 h of light and 8 dark (Merrill et al. 1990). Both of these situations are very different from the 16 sunrises and sunsets that the human crewmembers may experience (see Circadian Rhythms, Chap. 6).

Spaceflight Analog Studies: Rotating Cell Culture

When cultures of cells that grow well in suspension are placed in constant vertical rotation, the cells experience constant free-fall, similar to the spaceflight experience. Several devices have been developed that provide this kind of rotation to cell cultures, and the combination with microarrays or high-throughput gene-expression screening results in a powerful broad-spectrum tool for analysis of cellular metabolism. An analysis of two different Xenopus liver cell lines showed altered expression of several of the same genes (Ikuzawa and Asashima 2008). Of particular interest was down regulation of the SPARC gene that codes for osteonectin, an important protein in the extracellular matrix and thus one that is involved in cellular interactions (Ikuzawa and Asashima 2008). Unfortunately, the control for this study was a static culture, so the effect of motion in general cannot be separated from the effect of free-fall. The bioreactor developed by Nickerson and colleagues includes a sister culture that was placed in constant horizontal rotation to control for the effect of motion (Nickerson et al. 2000).

Spaceflight Analog Studies: Rat Tail Suspension

Various enzymes of Phase I metabolism have been studied in tail-suspended rats. In rats suspended for 3 or 7 days, an increase in oxidative metabolism was shown whereas the volume of distribution was not affected (Brunner et al. 1995). Amounts of liver cytochrome P450 enzymes changed in the first few days of tail suspension, followed by a correction back to baseline levels. Changes were seen in CYP2C11, CYP2E1, CYP4A1, and in p-glycoprotein. No change was seen in CYP3A2 (Lu et al. 2002).

Rat-tail suspension was shown to have no effect on Phase II drug metabolism measured with acetaminophen (Brunner et al. 2000). Acetaminophen is metabolized primarily by conjugation to form glucuronide and sulfate conjugates, and is subsequently excreted in urine and bile; it is considered a good marker for Phase II metabolic function (Slattery et al. 1987).

In tail-suspended rats, antipyrine administration led to increased hepatic oxidative function (Brunner et al. 1995). This finding is in direct conflict with the decrease in cytochrome P450 content seen in flown rats (mentioned with no citation in the *Skylab 3* report of Brunner et al.) (Brunner et al. 1995). However, Carcenac's study (1999) comparing stimulated cGMP production in tail-suspended rats and those flown on a 17-day flight showed large differences, suggesting that this flight analog may not be the best for some signaling processes (Carcenac et al. 1999). The authors also suggested that differences in age or strain could explain the disparity (Carcenac et al. 1999). However, a morphological study by Racine's group also found a lack of correlation between tail-suspended rats and those that experienced spaceflight (Racine and Cormier 1992).

Excretion

Drugs are excreted mainly in the urine, either intact in the case of hydrophilic compounds, or after liver metabolism has made hydrophobic compounds more polar. A small fraction of orally administered drugs are excreted unchanged in the feces, and some are sent into the biliary system, whence they will be either excreted in the feces or reabsorbed into the circulation and excreted in the urine. Since sweat, saliva, and tears are ultrafiltrates of plasma, tiny amounts of circulating drugs and their metabolites may partition from plasma into these fluids. This makes salivary sampling for toxicological or pharmacokinetic testing possible in some cases, but this must be tested for each individual drug (Gilman et al. 1990).

Renal filtration accounts for most drug excretion; the contributions of intestine, saliva, sweat, breast milk, and lungs are small, except for lung exhalation of volatile anesthetics. The kidney's function is to ensure stability of the volume of extracellular fluid, as well as its pH and concentrations of ions and metabolic products. Substances whose concentrations become too high are excreted in the urine, and those whose concentrations are low are reabsorbed back into the circulation (Berne and Levy 1988). Drugs and their metabolites are excreted in the urine, with the rate of excretion mainly dependent on blood flow to the kidney.

Drug molecules and metabolites that are bound to plasma proteins are excluded from entering the urine space by glomerular filtration. Because of this, plasma protein binding plays a role in drug excretion. A number of clinically important drugs have chemical characteristics that are similar to those of endogenous compounds, and they are handled by the same mechanisms as those compounds. For example, penicillin, an organic acid, is secreted into the urine by the same transport system that removes uric acid from the blood. In healthy people, drug excretion is predictable, but in cases of kidney failure or lack of pH homeostasis, drug excretion may stop or even reverse (substances ordinarily excreted may be reabsorbed). This can lead to overdose (Gilman et al. 1990).

A search of the literature shows no studies focusing on drug excretion in microgravity or during spaceflight. However, some information about renal and hepatic function exists that may provide clues to potential spaceflight effects on drug metabolism.

Since the arrival of the Urine Processing Assembly (UPA) on the ISS, drugs and their metabolites that are excreted into the urine have become much more important. The urine collected by the UPA is concentrated, disinfected, and treated to remove organic compounds and salts, with the goal of producing the purest water possible. Treatment of crewmembers with medications inevitably results in the addition of drug molecules to the urine collection tank. Each of these compounds has its own chemical properties that could interfere with the expected function of the UPA. The physical property of osmolarity could also play a role—as more solute molecules are added to the system, the likelihood of precipitation increases. However, the UPA incorporates multiple treatments that use a variety of purification methods, and water quality monitoring is routinely conducted.

Spaceflight Evidence

In-flight measurements indicate a slight reduction in total body water for the first few days of flight (Leach et al. 1991). This would be expected to reduce renal blood flow and drug excretion.

Spaceflight Analog Studies: Bed Rest

In a head-down-tilt (7-day) study using lidocaine as a probe for blood flow, small elevations in drug clearance and renal blood flow were found during tilt compared to before or after tilt. Daily variability was high, with some days showing no elevation. However, a consistent decrease (by ~30%) in plasma concentration occurred over the entire head-down period (Saivin et al. 1995). Whether this phenomenon occurs in spaceflight has not been tested.

Metabolism and Excretion Summary

Just as with drug absorption, in a worst-case scenario, if drug metabolism increased dramatically, it would be as if no medication had been given, leaving the original medical complaint essentially untreated. Alternatively, if metabolism slowed, plasma concentration of the drug would be higher than anticipated, leading to increased likelihood and severity of side effects, and possibly toxicity. This is complicated by the possibility that not all enzyme systems are equally affected by spaceflight and by the likelihood of drug interaction at the level of hepatic metabolism (Gilman et al. 1990).

As yet, no systematic effort has been made to examine the enzymes that metabolize drugs in spaceflight. The existing spaceflight evidence is opinion-based (Category IV), although in (non-established) ground models evidence from randomized, controlled studies (Category I) exists. Since it is feasible to house small mammals on board aircraft and the ISS, efforts to make the animals' environment more similar to that of their human counterparts would prove useful in resolving some of the conflicting reports in the literature. Data are available for only a few enzymes and range over a wide variety of flight analogs and other varied conditions. Knowledge of how and by which enzymes a drug is metabolized allows prediction of whether the compound may cause drug–drug interactions or be susceptible to marked individual variation in metabolism due to genetic polymorphisms. This is an area where use of modern molecular biological techniques (high-throughput gene-expression arrays, microarrays, difference gel electrophoresis, or DIGE, proteomics) could efficiently narrow the list of candidate genes and enzymes down to those affected by spaceflight (or a particular analog). More detailed experiments on these identified genes and enzymes could then be performed, especially those involving the enzymes that metabolize the drugs used in spaceflight.

The spaceflight evidence for this topic is observational (Category IV). Anecdotal reports from previous spaceflights have not indicated significant alterations in drug excretion. Preliminary studies in bed rest are descriptive (Category III).

Chapter 5
Central Nervous System

A great many of the drugs used in terrestrial medicine are given to treat central nervous system (CNS) symptoms (44% of all prescriptions in 2004; Agency for Healthcare Research and Quality 2006), and most drugs used for other reasons carry the risk of CNS side effects (Gilman et al. 1990). CNS drug use in space has not been any less frequent than on Earth, and this is not expected to change in the future. In fact, the two most common complaints that lead to medication use during space-flight are CNS-related: space motion sickness (SMS) and difficulties sleeping, which together account for 92% of medications used in space (Putcha et al. 1999). Significant dissatisfaction has been reported with medications used to treat both sleep trouble and SMS, but few reports exist of issues with other CNS-active thera-peutics during spaceflight, so this discussion will be limited to the reported areas of concern.

Sleep

Sleep is a special kind of unconsciousness, but unlike a comatose individual, a sleeping individual can be roused by sensory stimuli. Furthermore, selective atten-tion is granted to certain sensory stimuli, while others are ignored. For instance, most of us can sleep through a loud thunderstorm or traffic noise, but will awaken easily to the sounds of a fussy child. People use this "call-screening" feature with-out being aware of it. Also, there is a natural drive both to enter and to exit the state of sleep on a regular basis, usually in synchrony with circadian rhythms and envi-ronmental signals such as the daily light–dark cycle (Srinivasan et al. 2008).

The drive to sleep is known to all of us, but physiologically defining sleep, its functions, and its purposes have proven difficult for researchers. Nevertheless, the effects of sleep restriction or deprivation are well documented, and it is clear that sleep problems have significant costs in terms of human health and performance.

V.E. Wotring, *Space Pharmacology*, SpringerBriefs in Space Development,
DOI 10.1007/978-1-4614-3396-5_5, © Virginia E. Wotring 2012

Why Sleep Is Physiologically Required

Currently, there are several different theories to explain the purpose of sleep. The restorative theory says that sleep is needed for the removal of metabolic waste products accumulated by the active brain, regeneration of energy supplies, and synthesis of molecules that may have been depleted by the day's activities, such as signaling molecules (membrane receptors, transporters), molecules involved in energy production (ATP, electron transport enzymes), and many other synthetic enzymes (West 1969; Oswald 1976; Mackiewicz et al. 2007). The learning and memory theory says that during sleep time, the pattern of the day's activity of synaptic connections is rehearsed (which is experienced as dreams), consolidating and strengthening the day's memories into a more preserved or retrievable memory state. Evidence exists that learning depends particularly on rapid-eye-movement (REM) sleep, the sleep phase when dreaming occurs (Karni et al. 1994). The recent report by Vecsey et al. shows evidence that sleep deprivation prevents occurrence of the cAMP-dependent long-term potentiation that is normally seen and is thought to play a role in learning and memory (Vecsey et al. 2009). The recently rediscovered evolution theory of sleep function says that since most multicellular organisms sleep and each species has its own sleep pattern, the animals that slept must have had an evolutionary advantage over others that didn't (Webb 1974; Siegel 2009). Sleep may have come about to encourage an animal to remain hidden and quiet during times when its predators were active, or to conserve energy during times when it was less likely to encounter its prey (Siegel 2009).

Sleep Stages

Electrical measuring devices such as electroencephalography (EEG) have allowed sleep states to be more completely characterized by analyzing the summed neuronal activity that can be measured noninvasively with surface electrodes. These kinds of recordings have enabled researchers to identify five different kinds of sleep, based on the amplitudes and frequencies seen with EEG.

In stages 1–4, breathing and pulse slow and muscle tone relaxes, and the individual becomes more difficult to rouse. The EEG shows a progressive increase in waveform amplitude and reduction in frequency from stage 1 to stage 4. After about 30 min of stage 4 sleep, the sleeper generally moves back to stage 2, and then enters a stage of rapid eye movements, irregular heartbeat and respiration, and dreams, called REM sleep. The EEG of this REM sleep is very similar to that seen while subjects are awake. Repeating the cycle four or five times per night is typical (Rechtscheffen and Kales 1968). It is now clear that humans require sufficient time in both REM and non-REM sleep to feel rested and perform well. However, attributing specific physiological roles to these sleep stages has proven difficult.

Circadian Rhythms

It is also clear, however, that sleep is linked to biological rhythms (see Sack et al. 2007), for an excellent review of circadian rhythms and sleep disorders. Normal circadian rhythms are seen not only in fatigue and alertness, but also in body temperature, blood pressure (Agarwal 2010), hormone concentrations, protein synthesis, and other physiological variables. In the absence of external cues, these cycles tend to free run with a period of about 25 h, but the daily cycle of light and dark usually sets the period (Banks and Dinges 2007). It is thought that the requirement for environmental input into the system is adaptive in that it ensures that the individual's rhythms are appropriately synchronized with their local environment (Czeisler and Gooley 2007).

Humans have peaks in fatigue between 3 am and 5 am and again between 3 pm and 5 pm (Van Dongen and Dinges 2005). Most healthy individuals find it difficult to fall asleep at times of day not typical for them, which is especially problematic for travelers and shift workers (Srinivasan et al. 2008). It has also been noted that people who experience little outdoor light suffer from circadian rhythm misalignment and related insomnia, depression, and other disorders; indoor lighting tends to be too dim as well as too yellow for optimal circadian stimulation (Fig. 5.1) (Turner and Mainster 2008). Timed bright-light exposure has been successfully used to help crewmembers adjust to new time zones or night shift work (Whitson et al. 1995). The mechanistic details of this have not yet been elucidated, but light exposure in mammals leads to production of all-trans retinoic acid in the retina, which can then influence gene expression in the liver and affect metabolism (Pang et al. 2008).

Spaceflight Evidence

Sleep in spaceflight is similar to sleep deprivation. Sleep durations during missions in space are shorter than the 8 h recommended by NASA; in fact, crew sleep times average about 6 h (Nicholson 1972; Frost et al. 1975; Gundel et al. 1993; Gundel et al. 1997; Monk et al. 1998; Elliott et al. 2001). Clinically, this would be considered chronic sleep deprivation. However, it has been suggested that the body may actually require less sleep in a reduced-gravity environment. An examination of breathing during sleep has shown that in spaceflight, there is a marked reduction of respiration-related sleep disruptions such as apnea and snoring (Elliott et al. 2001). Therefore, it is possible that sleep quality is increased in space (Dinges 2001). However, since crewmembers routinely report difficulties sleeping as well as feelings of fatigue (Kelly et al. 2005), this does not seem likely.

The light environment experienced by crews is very different from what occurs on the ground. Normally people experience relative darkness (less than 3 lux) while they sleep, followed by dawn (3 lux) (Schlyter 2006), and then some mixture of daylight (ranging from 10,000 lux on a cloudy day to more than 30,000 lux in direct

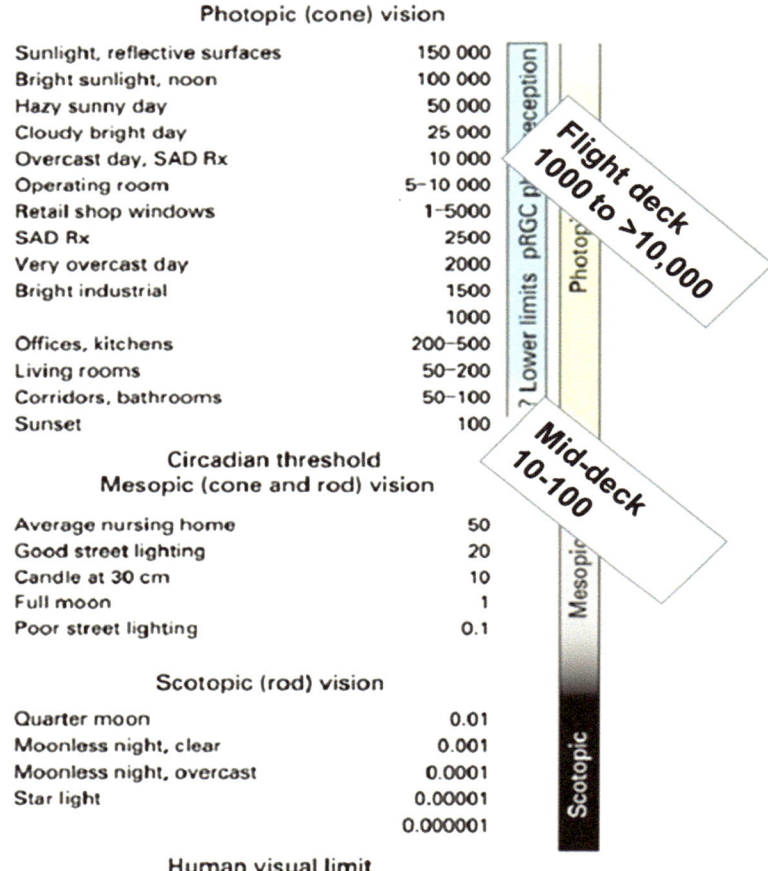

Illuminance (lux)
Photopic (cone) vision

Sunlight, reflective surfaces	150 000
Bright sunlight, noon	100 000
Hazy sunny day	50 000
Cloudy bright day	25 000
Overcast day, SAD Rx	10 000
Operating room	5–10 000
Retail shop windows	1–5000
SAD Rx	2500
Very overcast day	2000
Bright industrial	1500
	1000
Offices, kitchens	200–500
Living rooms	50–200
Corridors, bathrooms	50–100
Sunset	100

Circadian threshold
Mesopic (cone and rod) vision

Average nursing home	50
Good street lighting	20
Candle at 30 cm	10
Full moon	1
Poor street lighting	0.1

Scotopic (rod) vision

Quarter moon	0.01
Moonless night, clear	0.001
Moonless night, overcast	0.0001
Star light	0.00001
	0.000001

Human visual limit

Fig. 5.1 Light levels in various environmental circumstances in lux (Turner and Mainster 2008). Used with permission, BMJ Publishing Group Ltd

sun) and indoor artificial light (100–500 lux). Figure 5.1 shows lux measurements of typical situations for comparison. On the space shuttle, for example, typical indoor ambient lighting on the middeck and in Spacelab is dimmer than most terrestrial indoor lighting (between 10 and 100 lux), whereas the flight deck, with its large windows to the outside, has continual 90-min cycles with highs of 1,000 lux (sometimes almost 100,000) and lows of about 1 lux (Dijk et al. 2001). This is not unlike the cyclical paradigm of 15 min at 10,000 lux followed by 60 min at less than 3 lux, which was found to have phase-resetting properties similar to those of the same total time period of 10,000 lux (Rimmer et al. 2000). Thus it seems possible that the light environment outside the craft could be interpreted by the human brain as a continuous circadian phase-resetting signal, and could explain some or all of the sleep difficulties experienced in spaceflight.

What Happens in Sleep Deprivation: Performance Deficits in General

In general, sleep deprivation causes a reduction in performance on mundane, repetitive tasks while affecting performance in emergencies to a much lesser extent. Unfortunately, many of us routinely perform very critical tasks that are potentially dangerous, such as driving cars, giving medications to patients, or conducting security screenings. This is how sleep deprivation gets into the news: a bad traffic accident, the Exxon Valdez oil spill, a medical mistake. Sleep deprivation is implicated in more than 100,000 car accidents per year in the United States (Knipling et al. 1995), and people who reduce their nightly sleep by about or more than 2 h per night are twice as likely as others to die from any cause (Hunter 2008).

In an effort to determine the functions of sleep by examining deficits that occur in its absence or reduction, the performance of sleep-deprived individuals on various types of cognitive and motor tasks has been measured. The literature on this subject is vast and includes retrospective examination of accidents (traffic, industrial, and others), including the sleep history of individuals involved, as well as laboratory experiments in which performance was measured on tasks specifically designed to separate cognitive elements from motor elements in situations of known sleep history. Attempts also have been made to differentiate between vigilance, or attention, and neuronal processing time, as in the psychomotor vigilance test (PVT) (Drummond et al. 2005; Van Dongen and Dinges 2005; Lim and Dinges 2008; Lim et al. 2010). The digit substitution test, another commonly used laboratory test, is a more complex cognitive and memory task.

A relatively minor level of sleep reduction (only 2 h reduction per night for 1 week) has been shown to cause significant performance impairment as measured by psychomotor vigilance test, digit substitution test, and driving simulators (Banks and Dinges 2007; Lim and Dinges 2008). Jet lag and shift work cause similar performance deficits, even in people accustomed to constantly changing schedules (Cho et al. 2000; Basner et al. 2008).

Sleep deprivation is detrimental not only to attention and task performance but also to many of the body's physiological systems (Mullington et al. 2009). Tests of basic physiology have shown that sleep deprivation alters immune system function (see the excellent review by Simpson and Dinges (2007)), raises blood pressure (Tochikubo et al. 1996; Kato et al. 2000; Ogawa et al. 2003; Meier-Ewert et al. 2004), and increases mediators of inflammation (Kuhn et al. 1969; Dinges et al. 1994; Boyum et al. 1996). Recent reports linking sleep deprivation to obesity (Laposky et al. 2008a, b) emphasize how fundamentally important proper sleep is for normal physiological function. Sleep difficulties have also been linked to development of post-traumatic stress disorder, possibly due to the lack of REM-phase memory consolidation that should occur in sleep but is blocked by trauma-induced sleep difficulties (Germain et al. 2008).

Performance Deficits Specific for Spaceflight

Although crewmembers have been reporting difficulties sleeping since the beginning of space exploration, and they have used medications accordingly, few objective measures have been made of performance in flight. Small performance decrements have been demonstrated on a number of recognition tasks and the digit substitution test (Kelly et al. 2005). For the first 90 days of spaceflight, measurements of physiological parameters under circadian control (body temperature and alertness) have been shown to remain constant (in a single individual), but they began to deviate from 24-h cycles over the next 2 months of flight (Monk et al. 2001). It is not yet known whether we can generalize from this result.

It seems that total sleep time per 24-h period is the critical factor for maintaining crew performance. In a study of split sleep schedules compared to typical schedules with single sleep episodes, no performance differences were noted when sleep time was split into two shorter sessions (Mollicone et al. 2008).

Therapies for Sleep Difficulties

Throughout our history, humans have probably always experienced occasional problems sleeping. The earliest treatment available was likely ethanol, which was administered by physicians until the early twentieth century and is still used by individuals treating themselves (Katzung 2007). The first barbiturates became available at the beginning of the twentieth century, but had serious side effects, including significant next-day residual lethargy and confusion (so-called "hangover effects" due to the 10–60-h half-lives of these drugs) as well as frequent overdose and dependency problems (Gilman et al. 1990; Hindmarch and Fairweather 1994). By the 1960s these untoward effects were becoming a concern to prescribers, which helped the new benzodiazepines become the drugs of choice for insomnia and anxiety. Chlordiazepoxide (Librium®) and diazepam (Valium®) seemed safe by comparison with earlier alternatives, and both physicians and patients felt comfortable with their routine use (Walsh and Engelhardt 1992). Since then, many of the negative effects of barbiturates have been found to be associated also with benzodiazepines, which is not surprising because both act as agonists (activators) of neuronal inhibitory GABA receptors (Gilman et al. 1990).

Now a new class of non-benzodiazepine sleep medications has become available. Both zaleplon (Sonata®) and zolpidem (Ambien®) have been shown to effectively promote sleep by decreasing sleep induction time and reducing awakenings during the night (Berlin et al. 1993; Scharf et al. 1994; Beaumont et al. 2007).

Zolpidem seems to have little effect on slow-wave sleep but decreases REM sleep, whereas zaleplon does not seem to affect any of the sleep phases (Katzung 2007). Both zaleplon and zolpidem improve sleep duration and quality with minimal performance effects on the morning after use (Hindmarch and Fairweather 1994; Walsh et al. 2000; Beaumont et al. 2007). Differences between zaleplon and zolpidem have been noted in studies of early awakening. Residual sedation was not found 2–4 h after zaleplon dosing, but was seen up to 7 h after zolpidem dosing (Berlin et al. 1993; Danjou et al. 1999; Hindmarch et al. 2001b; Patat et al. 2001; Zammit et al. 2006), a finding that is consistent with their respective half-lives (zaleplon ~1 h; zolpidem ~3 h (Gilman et al. 1990)). These non-benzodiazepine sleep aids are becoming popular and are perceived as safe and not habit-forming (Greenblatt et al. 1998a; Siriwardena et al. 2006; Siriwardena et al. 2008), although current prescribing guidelines call for investigation of insomnia cause and use of behavioral therapies, as well as a time limit to prescription sleep aid use (Schutte-Rodin et al. 2008). In time, new potential therapies may become available as more is learned about orexin receptors and their role in arousal and sleep (Mieda and Sakurai 2009; Nunez et al. 2009; Coleman et al. 2004).

Melatonin is a hormone whose production by the pineal gland is suppressed by light. It is not used as a sleep aid per se, but is used to reset circadian rhythms so that sleep at an appropriate time becomes easier (Hardeland et al. 2008; Srinivasan et al. 2009). Exogenous dosing during the newly desired "night" has been shown to help reset circadian rhythms in people with jet lag and in blind people who show no endogenous rhythm (Arendt et al. 1986; Arendt et al. 1986; Folkard et al. 1990). A high-dose protocol did not noticeably improve sleep when used in flight (Dijk et al. 2001), but given that there has been no consensus in the literature as to the best formulation, dose, or dosing schedule (Sack et al. 1997), it is possible that the dose used in this flight study was not optimal. Circadin™ is a prolonged released formulation of melatonin in clinical trials in the United States and approved for use in Europe (Srinivasan et al. 2009). Recently, a more potent agonist of melatonin receptors called ramelteon has been shown to advance the phase of circadian rhythms with no significant side effects (Richardson et al. 2008; Miyamoto 2009). Either or both of these may prove to be more useful than melatonin.

Both zolpidem and zaleplon have been used as sleep aids in flight. Several other drugs have been used in the past: the benzodiazepines temazepam (Restoril®), flurazepam (Dalmane®), and triazolam (Halcion®) as well as the sedating antihistamines promethazine (Phenergan®) and diphenhydramine (Benadryl®) (Taddeo and Armstrong 2008).

Risks of Sleep Aid Use Unique to Missions

When sleep aid medications are prescribed terrestrially, the patient is told to take the drug upon retiring and only when at least 8 h will be available for sleeping. If other activities might need to be performed in the 8-h period after dosing, especially activities such as driving, the drug should not be taken (FDA 2008). In spaceflight, however, this may not be feasible. There is always the possibility of an emergency that requires the crew to be awakened early. And it is possible that they will be called upon to perform mission-critical and/or life-saving tasks sooner than 8 h after having taken a sleep aid medication. This possibility is unlikely, but the potential consequences of impaired performance are devastatingly high.

Potential Countermeasure for Sleep Aid Risks

Several strategies could be employed to minimize this risk. One would be to refrain from use of sleep aid medications. Of course, this solution could leave the issue of sleep loss and associated fatigue unresolved, unless crewmember sleep quality and duration could be adequately improved by behavioral approaches (for example, using cues like timed bright-light exposure to reset circadian rhythms (Czeisler et al. 1991; Dijk et al. 2001; Fucci et al. 2005)) or environmental means (such as improving noise and light distractions in sleeping quarters at night, Becker and Sattar 2009).

Another option is to use a pharmacological tool to reset circadian rhythms. Melatonin and ramelteon have been used to treat shift workers, international flight crews, and blind people. Moderate improvements in sleep times, sleepiness at inappropriate times, and fatigue symptoms have been shown to occur, and no significant side effects have been reported (Petrie et al. 1993; Lewy et al. 2001). Recently, the new melatonin agonist/serotonin 2C antagonist agomelatine has been shown to phase-shift rats, with few side effects, and merits further attention as studies in humans progress (Descamps et al. 2009). If indeed the crew sleep difficulties are associated with disruption of circadian rhythms, melatonin or one of its agonists would seem a practical therapy to shift the circadian clock to what is desired, and would likely not impair performance (Erman et al. 2006; Otmani et al. 2008) or cause problems in early-awakening situations.

Another possible countermeasure is to develop a fast-acting antidote to the sedating sleep aids used in flight. A general stimulant drug (dextroamphetamine) has been carried on board, and although its use is a standard treatment for fatigue in flight medicine (Bonnet et al. 2005) and it would certainly be an excellent second-choice treatment in the emergency awakening scenario, there may be better drugs for this situation. Although the use of dextroamphetamine in fatigue has been well studied (Caldwell et al. 2000), its use as an emergency countermeasure for sleep medication use has not. Residual detrimental effects on memory would not be expected to be reversed by dextroamphetamine use (Ko and Evenden 2009), and in fact, at higher doses (or after several doses) dextroamphetamine may reduce memory consolidation and impair learning.

Amphetamine has also been shown to reduce the ability to search a scene for visual details that may be pertinent to the task at hand (Kennedy et al. 1990) and to impair the ability to discriminate important information from irrelevant input (Swerdlow and Geyer 1998). Furthermore, since dextroamphetamine has a long half-life, it may be 2 or 3 days after dosing before a return to normal sleep pattern would become possible (Caldwell et al. 2000; Queckenberg et al. 2009). Amphetamine also carries a well-known potential for abuse. Another possible antidote is caffeine, a familiar stimulant that improves performance and is already carried on board in beverage form. However, it is not very effective with repeated use over a prolonged period, and its diuretic side effect may be an operational issue (Hindmarch et al. 2000; Van Dongen et al. 2001).

Modafinil is another stimulant (Makris et al. 2007). It probably acts by reducing GABA release, rather than affecting dopamine uptake like amphetamine (Ferraro et al. 1997), although limited evidence also exists for several additional mechanisms of action (Wesensten 2006). Modafinil improves vigilance and alertness, and causes few side effects (Dinges et al. 2006; Grady et al. 2010), although there are risks of cardiac and CNS problems in overdose (Spiller et al. 2009). Both caffeine and modafinil have been shown to reduce willingness to engage in risky behaviors when given to sleep-deprived individuals (Killgore et al. 2008). Intentionally providing crewmembers with drugs that enhance timidity and caution may not be advisable.

Most sleep aids increase activity of the type A γ-aminobutyric acid receptor complex ($GABA_AR$) expressed by many neurons in inhibitory signaling pathways (Gilman et al. 1990; Katzung 2007). Pharmaceuticals are available that interact with the same site on the $GABA_AR$, but have the opposite effect on receptor function. They are not regularly used in clinical practice to counteract sleep aids, but they have been used successfully in cases of benzodiazepine overdose or adverse reactions (Misaki et al. 1997; da Silva et al. 2008). Flumazenil has been shown to cause a recovery from coma and sleepiness after zolpidem administration, and also from drug-induced memory deficits (Wesensten et al. 1995; da Silva et al. 2008). Use of flumazenil has not been reported for an early-awakening scenario; the current literature has no information about the degree of performance improvements (motor or cognitive) after awakenings earlier than 8 h.

Central Nervous System Summary

The spaceflight evidence falls into Category II (controlled studies), and evidence from a number of ground studies is in Category I (controlled and randomized studies). Given the frequency of crew complaints about sleep, optimizing a plan for improving crew sleep should be a very high priority. This plan should include behavioral elements as well as pharmaceuticals. Sleep aids act on the same receptor, and it is known that overactivation of this receptor leads to respiratory depression. Ambien/Sonata is a potentially hazardous mix, and the use of these sleep aids in the same night should be tested on the ground. Combination with other drugs used in flight (promethazine, for example) should also be ground tested to prevent polypharmacy problems. A sleep aid antidote should be made available in case of emergency during sleep periods.

Chapter 6
Cardiovascular System

The cardiovascular (CV) system is responsible for the circulation of blood to all body tissues. The absence of Earth's gravity removes a significant force against which our bodies usually have to work; much of our normal exertion goes toward remaining upright and balanced in gravity (Sockol et al. 2007). During spaceflight, this need is removed, and the muscles of the body, including the heart, don't have to work so hard. In general the CV system adapts well to spaceflight, but some problems have been encountered in our space travel history. Just as on Earth, the CV system responds well to exercise and not well to extended breaks from exercise. Orthostatic intolerance and cardiovascular deconditioning are problems for many of our returning crewmembers. These risks have been recently reviewed in the Risk Reports from the Cardiovascular Discipline (Platts 2008b); only recent findings and those particular to pharmacological therapies will be presented here.

Upright bipedal humans have some adaptations that, on Earth, tend to maintain continuous perfusion and pressure throughout the body, despite a 5–7-foot difference in hydrostatic pressure. When gravity's force is removed, these adaptations can cause problems.

For instance, in humans capillary basement membrane thickness is the same in all body parts at birth, but it gets thicker in the dependent portions of the body as development continues, until it is about twice as thick at the body's lowest points as at its highest (1894 Å in the gastrocnemius compared to ~900 Å in the abdomen or the pectoral muscle) (Williamson et al. 1971). This means that if pressures are equalized on all parts of the body, as they are in microgravity, the capillaries of the head will experience higher pressures than normal (Fig. 3.1) (Hargens and Watenpaugh 1996) When higher pressure is coupled with a relatively thin basement membrane, the result is more transcapillary leakage than would occur in 1 G. This feature of cardiovascular anatomy resulting from its adaptation to 1 G may play a role in papilledema and headaches associated with spaceflight. In fact, differences in transcapillary fluid movement have been noted between bed-rested subjects who

V.E. Wotring, *Space Pharmacology*, SpringerBriefs in Space Development, DOI 10.1007/978-1-4614-3396-5_6, © Virginia E. Wotring 2012

Fig. 6.1 Plasma renin on high-sodium (*control*) and low-sodium (*sodium-restricted*) diets (Davrath et al. 1999). Used with permission, Aerospace Medical Association

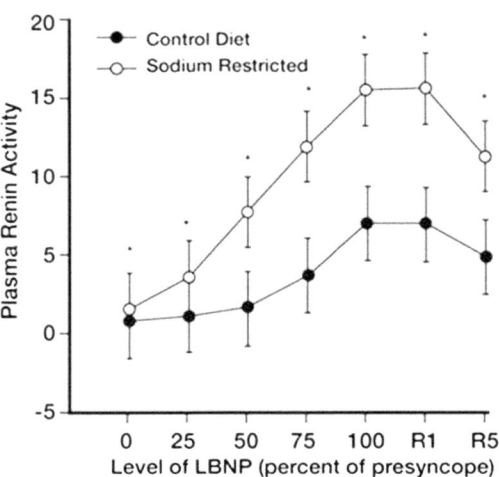

experience orthostatic intolerance upon tilt testing and those who are resistant, suggesting that people with greater amounts of fluid movement across capillaries are more likely to experience frank syncope (fainting) (Hildebrandt et al. 1994). Differences in vessel wall thickness and reactivity have also been seen before and during bed rest in the lower extremities, but not in the upper extremities (Platts et al. 2009a).

The complicated cardiovascular regulatory system shows many changes in spaceflight (Charles and Bungo 1991; Platts 2008a, b). Some of these are changes in the EKG (Mulvagh et al. 1991; Lathers et al. 1993). See the excellent reviews by Mano (Mano and Iwase 2003; Mano 2005) for a discussion of the role of the nervous system in postflight orthostatic intolerance (POI).

Cardiac contractility decreased in a spaceflight analog cell culture system, likely through a nitric oxide pathway (Xiong et al. 2003). Atrial naturietic peptide (ANP) is secreted by the heart in response to sympathetic drive, high sodium, or atrial distension (Guyton and Hall 2006). ANP has been shown to increase with time spent in head-down-tilt bed rest, on a scale of minutes (Grundy et al. 1991), probably via atrial stretching caused by fluid redistribution. However, in ground tests, ANP treatment reduces capillary filtration and keeps more fluid in the capillaries, which would tend to prevent orthostatic problems (Watenpaugh et al. 1995).

The sodium levels of the current ISS diet are complicating the interpretation of much CV research, although steps are being taken to remedy this. A high-sodium diet suppresses sympathetic nervous system activity and the renin-angiotensin system (Fig. 6.1) (Davrath et al. 1999). This raises the possibility that a high-salt diet could prevent the increase in sympathetic output that is required to compensate for postflight orthostatic hypotension.

Spaceflight Evidence

Cardiac Atrophy

Cardiac atrophy has been noted in some returning astronauts. It has not been associated with any adverse events, but it has been suggested that atrophy contributes to orthostatic intolerance (Levine et al. 1997; Platts 2008b). Left ventricular mass has been shown to drop by 12% during short flights ($n=4$ astronauts) (Perhonen et al. 2001). This topic is the subject of an ongoing flight study using ISS crewmembers that is projected to be complete in 2013. Preliminary results show greater atrophy after long-duration flights, and this atrophy does not recover by the third day after landing (Platts 2008a). In a ground study, hypovolemia was induced and was found to be correlated with reduction in left ventricular mass, with the conclusion that the apparent cardiac atrophy in space is directly related to fluid shifting and probably not a matter of significant concern on its own (Summers et al. 2005).

Cardiac Arrhythmias

Cardiac arrhythmias have been reported to occur in crewmembers during long-duration flight, but these have not been associated with any adverse events (Jennings et al.; Smith et al. 1976; Rossum et al. 1997; Fritsch-Yelle et al. 1998; Migeotte et al. 2003; Platts 2008a). The physical demands of extravehicular activities have been ruled out since they did not correlate well in time with the arrhythmias (Rossum et al. 1997). It is generally presumed that these rare occasions were likely associated with previously undetected cardiac disease, but it remains possible that some aspect of long-duration spaceflight may play a role (D'Aunno et al. 2003). This topic was recently covered by the Evidence Report from the Cardiovascular Discipline (Platts 2008b). However, it should be noted that many pharmaceuticals block the cardiac HERG potassium channel, which prolongs the QT interval (Keller et al.; Gilman et al. 1990), and this could manifest clinically as arrhythmia. It is possible that the arrhythmia incidents seen during flight were related to pharmaceuticals used during missions.

Postflight Orthostatic Intolerance

Postflight orthostatic intolerance (POI) prevents astronauts from immediately resuming their typical terrestrial activities and carries the risk of severely impairing their ability to perform in a landing emergency (Broskey and Sharp 2007). In postflight tests, 9 of 14 STS (shuttle) astronauts were unable to tolerate 10 min of standing (Buckey et al. 1996). Among STS astronauts, 100% of the women and 20% of the men experienced POI (Waters et al. 2002; Platts 2008b).

The mechanism of POI is not completely understood, although significant improvements in understanding have been in made in the past few years. People who have experienced POI once are likely to have repeat occurrences on subsequent flights (Martin and Meck 2004). It has been noted that most people experiencing POI develop it over a few minutes of standing, which would be consistent with a slow mechanism of action, such as capillary infiltration from the vasculature to the tissues (Broskey and Sharp 2007) or venous pooling (Verheyden et al. 2009). A lack of compensation in the peripheral vasculature by the sympathetic nervous system has been demonstrated in returning astronauts who fail to finish tilt tests (Buckey et al. 1996). Preflight lack of α-receptor activation by norepinephrine was clearly shown in 9 presyncopal crewmembers compared to 13 non-presyncopal counterparts, and may be responsible for individual susceptibility to orthostatic issues (Meck et al. 2004). Decreased vagal tone has been noted in returning crewmembers (Migeotte et al. 2003).

It is also interesting to note that for the single elderly astronaut who has been tested, most CV variables were the same before and after flight, but this individual showed a much larger release of norepinephrine than younger counterparts and showed no orthostatic intolerance upon landing (Rossum et al. 2001). A decrease in central integration of baroreceptor input may explain these findings, but this notion remains unconfirmed. These are all clues to a mechanism, but the complete story has yet to be unraveled.

As in terrestrial medicine, plasma volume is clearly associated with orthostatic tolerance. Various fluid-loading and pressure-garment strategies have been and continue to be evaluated as useful countermeasures. Salt-and-fluid loading increases plasma volume, is thought to help reduce OI, and is implemented in current operations (Bungo et al. 1985; Frey et al. 1991). However, it alone has not solved the problem.

Non-pharmaceutical measures have also been investigated. Compression garments reduce POI (Platts et al. 2009b) and continue to be studied as countermeasures. Anecdotal reports exist of behavioral maneuvers, such as the tensing of leg muscles, being used to prevent POI, but these do not seem to have been systematically tested (Gisolf et al. 2005).

Several pharmacological countermeasures to POI have also been tested. Trials were conducted with fludrocortisone, since it had been shown to increase plasma volume in terrestrial orthostatic hypotension (OH) patients (Vernikos et al. 1991). A single 0.3-mg dose hours before landing did protect plasma volume in returning astronauts, but it did not alter the rate of presyncope (Shi et al. 2004). Although effects of only the single 0.3-mg dose were studied, fludrocortisone was dropped from further study because of untoward effects (Shi et al. 2004). It is possible that a different dose or dosing schedule would prove effective, if plasma volume actually plays a role in OH. These data from the Shi study (Shi et al. 2004) suggest a limited role of volume in OH, but are not conclusive.

It has also been suggested that a reduction in normal mobilization of splanchic circulation may play a role in POI (Ray 2008) and that an effective countermeasure for crewmembers might involve mobilizing this pool of blood volume. Midodrine has also been tested as a countermeasure. It was shown to have no significant side effects in non-symptomatic returning crewmembers (Platts et al. 2006b), although it has very recently been called into question by the FDA for insufficient proof of efficacy (FDA 2010). In a bed rest study, it did seem to reduce POI (Ramsdell et al. 2001); however, significant side effects were observed when it was used in combination with promethazine (PMZ) (Platts et al. 2006a), currently the most effective treatment for space motion sickness (SMS), and often taken around the time of landing.

PMZ, routinely used to prevent and treat space motion sickness, has recently been shown to increase the likelihood of orthostatic hypotension (Shi et al. 2010). Other PMZ side effects were discussed in the SMS section above, but it's possible contribution to OH is especially troublesome, particularly if landing logistics require additional crew activity. Ideally, NASA would switch to a different SMS therapy, but at the current time, the only other drug found to be effective for SMS is scopolamine, which is less effective and more likely to cause side effects. Until a more promising anti-motion sickness drug becomes available for testing as an SMS therapy, it seems that the best solution is to treat the OH.

Spaceflight Analog Studies

Cardiovascular issues were among the first tested on crewmembers, and appropriateness of analogs has been reasonably well established (Pavy-Le Traon et al. 2007). Although many human studies are conducted in head-down-tilt bed rest, some use water immersion or dry immersion, and the results are not identical, although in general the agreement is good (Shiraishi et al. 2002; Waters et al. 2005). Some studies also use parabolic flight, with its brief alternations between high G and microgravity. In parabolic flight, a constriction of cerebral arteries has been shown to be linked with orthostatic intolerance (Serrador et al. 2000), but the relationship of this extremely brief microgravity exposure to the spaceflight experience has not been firmly established. It has also been shown that changes in head position during parabolas altered the measurements of peripheral blood flow (Herault et al. 2002), which may confound results from this analog.

The mechanistic studies that are possible in animal research have shown that the nervous system clearly plays a role in POI. Orthostatic hypotension is increased by sectioning the vestibular nerve (Yates et al. 1998). In hind limb-suspended rats, reduced sensitivity of $\alpha 1$-adrenoreceptors has been shown, and this could explain the apparent lack of effect of norepinephrine (Sayet et al. 1995), although in bed rest (30 days), vascular responsiveness to exogenous norepinephrine was normal (Convertino et al. 1998). Midrodrine, an $\alpha 1$ agonist, has been shown to reduce OI after (16-day) bed rest (Ramsdell et al. 2001), but significant side effects have been noted when this drug is combined with the PMZ used to combat SMS, so this drug

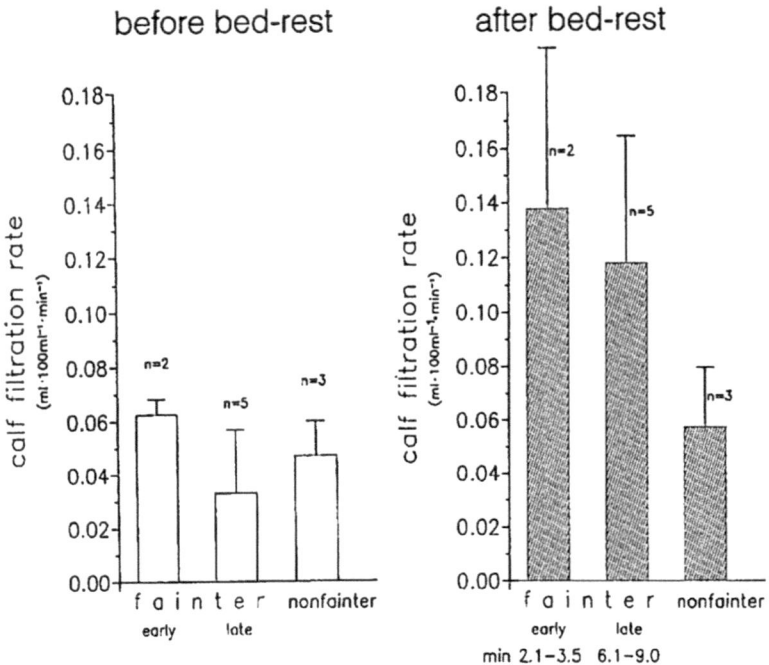

Fig. 6.2 Calf filtration rates of presyncopal and non-presyncopal subjects, before bed rest (*left*) and after (*right*) (Hildebrandt et al. 1994). Used with permission, Springer

combination is not used (Platts et al. 2006a; Shi et al. 2010). The possible role of nitric oxide in POI was implicated in tail-suspended rats. Although nitric oxide level was not affected by tail suspension, it increased significantly in the few hours immediately after the animals were returned to an upright position (Bayorh et al. 2001).

Fludrocortisone (a mineralocorticoid hormone that increases retention of salt and water at the kidney) has been shown to be more effective at preventing POI than saline loading in a (7-day) bed rest model (Vernikos et al. 1991; Vernikos and Convertino 1994) and has a long-established connection with plasma renin activity (Thompson et al. 1979). In the 1991 study dextroamphetamine was administered 1 h before the stand tests (Vernikos et al. 1991), which makes comparisons to other studies less straightforward. The "leakiness" of the peripheral vasculature has also been investigated in the bed rest model. Volume of the calf muscle was measured after 24 h of bed rest, and a significant difference was seen between the individuals who were tolerant of an orthostatic challenge and those who were not, as shown in Fig. 6.2 (Hildebrandt et al. 1994). ANP treatment reduces capillary filtration and keeps more fluid in the capillaries, which would tend to prevent orthostatic problems (Watenpaugh et al. 1995).

Cardiovascular Summary

The evidence for cardiovascular system changes during spaceflight falls under Category III, observational, with certain studies that involve interventions falling under Category II, controlled studies that use subjects as their own controls. The many bed rest studies discussed above are Category II. Unfortunately, there is disagreement among various research groups about the mechanisms of physiological changes, and it seems quite likely that additional data from launch and landing periods would clarify the subject.

Chapter 7
Gastrointestinal System

Little is known about the function of the gastrointestinal (GI) system in spaceflight or any changes that might occur in microgravity. Few studies have attempted to address this physiological system in spaceflight, but empirically, our years of spaceflight experience have shown that GI function in microgravity is not very different from GI function in 1 G. Crewmembers have been able to eat and drink without major problems, and when their intake is sufficient, they maintain body weight. Some unsettled questions remain about GI motility and absorption of medications, discussed above under "Absorption." But by far, the most significant GI complaint in spaceflight is motion sickness, which is not solely a GI problem—it has a significant CNS element (Muth 2006), which is discussed below.

Motion Sickness

It has been known for centuries that experiencing passive motion can affect the human body; Hippocrates noted this fact regarding seasickness in *The Nature of Man* about 2,400 years ago (Megighian and Martini 1980). The term *motion sickness* was first proposed in 1881 (Graybiel 1976), not long after the automated transportation of humans became commonplace, first by rail, then by automobile.

Some people are quite susceptible to motion-induced illness, whereas others are quite resistant, and the reason for this difference is not well understood. Space motion sickness (SMS) is the term for the suite of symptoms experienced by astronauts, and it is thought to be motion-induced. SMS includes nausea, pallor, cold sweating, and sometimes vomiting. About 70% of astronauts report SMS symptoms (Davis et al. 1988), particularly during periods of transition from Earth's gravity to microgravity and back again (Buckey 2006). The associated debilitation is considered serious and frequent enough so that operational demands on crewmembers are kept to a minimum on the first 3 days of flight.

V.E. Wotring, *Space Pharmacology*, SpringerBriefs in Space Development,
DOI 10.1007/978-1-4614-3396-5_7, © Virginia E. Wotring 2012

Table 7.1 The dimensions of the motion sickness questionnaire (Muth 2006). Used with permission, Elsevier

Gastrointestinal	Central	Peripheral	Sopite-related
Sick to my stomach	Faint	Sweaty	Annoyed/irritated
Queasy	Light-headed	Clammy/cold sweat	Drowsy
Nauseated	Disoriented	Hot/warm	Tired/fatigued
As if I may vomit	Dizzy, like I was spinning		Uneasy

Unique Aspects of Space Motion Sickness

SMS generally occurs during periods of environmental transition, in either the first few days of flight or the first few days back on Earth, or both (Buckey 2006). For most crewmembers, symptoms that are present during flight days 1–3 will disappear for the remainder of a flight (Buckey 2006). This indicates that SMS is not driven by passive motion alone, since the motion is constant over the entire duration of a flight. SMS symptoms are a little different from terrestrial motion sickness in that sweating and pallor may be absent or relatively minor symptoms, and also in that vomiting may occur suddenly with little warning (Table 7.1) (Ortega and Harm 2008). The vomitus appears unusual in that it is clear, and if it contains food, it is undigested food. SMS is also typified by lack of bowel sounds, an indication of low GI motility (Thornton et al. 1987). Sometimes SMS symptoms are limited to what is called the sopite syndrome: drowsiness, lack of initiative, lethargy, and apathy, with little to no stomach involvement (Graybiel and Knepton 1976). Clearly, nausea and vomiting are debilitating and mission-threatening, but the sopite syndrome also has a strongly negative effect on crew performance, bordering on or including signs of depression (Graybiel and Knepton 1976). Consequently, although nausea and vomiting are primary symptoms, SMS is considered a problem of the central nervous system (CNS), not of the GI tract.

Nausea and Vomiting

There is a large volume of literature on nausea and vomiting (N&V), and it is tempting to consider it a single disorder. However, motion-induced vomiting does not respond to the same therapies that are effective for vomiting induced by chemicals (such as cancer chemotherapy, pregnancy, or surgical anesthesia), suggesting that the mechanisms responsible for motion-induced illness are distinct from those that lead to other N&V syndromes. Some newer anti-vomiting drugs in the ondansetron family have revolutionized cancer chemotherapy, but have proven to be ineffective in treating motion-induced illness (Stott et al. 1989; Levine et al. 2000). This fact has not been widely noted among clinicians and researchers outside of NASA, and has led to some confusion regarding NASA's therapeutic choices for SMS.

In ground studies, training has been shown effective in prevention of only certain N&V syndromes. Anticipatory nausea (the nausea experienced when a person is returned to an environment that previously produced nausea) is reduced by training, whereas rotational nausea (caused by a spinning chair or centrifuge) is unaffected by training (Klosterhalfen et al. 2005). This also supports the notion that different kinds of nausea have different mechanisms and consequently require different therapies.

Furthermore, nausea and vomiting are thought of as inexorably linked—when nausea intensifies, it culminates in a bout of vomiting. However, some evidence from clinical practice indicates that nausea and vomiting may in fact be separate phenomena. The most notable support for this is the remarkable efficacy of $5HT_3$ antagonists such as ondansetron for prevention of vomiting, but many patients still experience nausea even though they have had relief from vomiting (Sanger and Andrews 2006). It has been proposed that nausea is produced by actions of the forebrain (in response to input from the brainstem), whereas vomiting is initiated in the brainstem with no forebrain contribution (Horn 2007). Different neurotransmitters and receptors are used for cellular communication in these different parts of the brain, which would explain the differences in drug efficacy.

The Mechanism of Space Motion Sickness

The most widely accepted theory of the SMS mechanism is that a mismatch between inputs from the vestibular system and those from the visual system causes disruption in higher brain centers that receive conflicting information. Each of these senses (visual and vestibular) delivers information to the CNS about body position and movement, and it is thought that when the CNS receives conflicting information, problems may arise (Reason and Brand 1975; Yates et al. 1998). Ordinarily, the information from the visual system regarding the scene in front of the eye (and thus the inferred position of the body in space), the information from foot Merkel disks that signal the pressure on the foot while standing on the floor, and the information from the body's proprioceptors and otoliths regarding body position all mesh to provide the CNS with different views of the same situation, as shown in Figs. 7.1 and 7.2. In contrast, in microgravity the otoliths, vestibular system, and skin pressure sensors may not signal anything to the CNS, or they may send signals that are incoherent and meaningless. This would result in an apparent mismatch when their inputs are compared to the input from the visual system. Limited evidence suggests that crewmembers experience a reduction in vestibular input as they adapt to spaceflight, which would tend to support this theory (Watt and Lefebvre 2003).

In the brain, the medullary vomiting center receives inputs from the nearby chemoreceptive trigger zone (mediated largely by D2 dopamine receptors and $5HT_3$ serotonin receptors), the vestibular apparatus (mediated by muscarinic and H1 histamine receptors), as well as peripheral afferents (Fig. 7.2). The drugs used therapeutically to reduce N&V may act on any or all of these receptor types (Katzung 2007). The role of acetylcholine receptors in motion sickness is supported by the

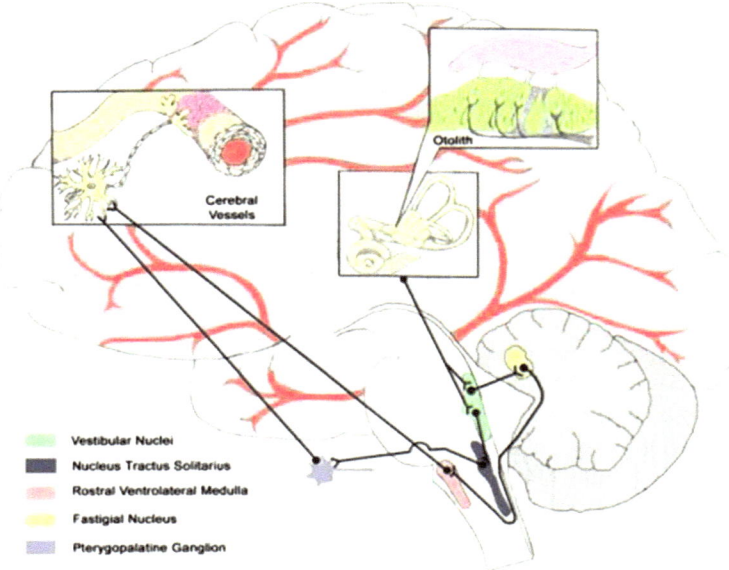

Fig. 7.1 Pathways connecting vestibular organs and cerebral vessels (Serrador et al. 2009). Open Access

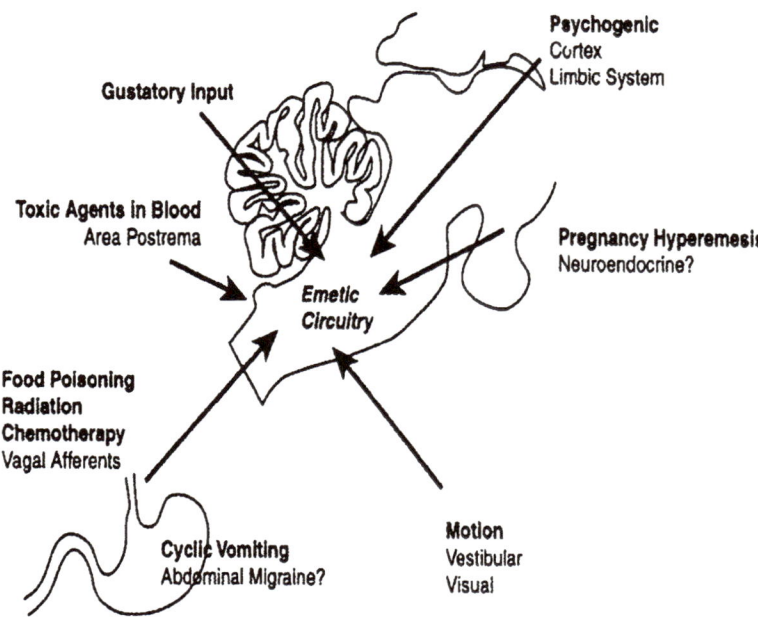

Fig. 7.2 The emetic circuitry in the hindbrain (Hornby 2001). Used with permission, Elsevier

fact that physostigmine, a cholinergic agonist, can be used to induce symptoms similar to those of motion sickness (Janowsky et al. 1984).

The sensory mismatch theory is by far the most widely accepted, although others exist. It has been proposed that an imbalance in labyrinthine cerebrospinal fluid (CSF) is the physiological trigger for SMS (Parker 1977). A properly functioning labyrinth organ has been shown to be required for ground-based motion sickness to develop, but this notion has not been directly tested in microgravity (Kennedy et al. 1968; Borison 1983). It has also been suggested that elevation of intracranial pressures may play a role in SMS, and although changes in cerebral perfusion have been demonstrated in ground centrifugation (Serrador et al. 2005), this theory also has not been directly tested (Lakin et al. 2007).

SMS Susceptibility

General trends have been noted regarding the nausea caused by cancer chemotherapy; susceptibility to terrestrial motion sickness correlates with increased likelihood and greater severity of nausea and vomiting during cancer chemotherapy (Morrow 1984). Unfortunately, there seems to be little relationship between susceptibility to terrestrial motion sickness (air, sea, train, automobile) and the likelihood of experiencing SMS (Heer and Paloski 2006). Previous spaceflight experience does not seem to affect the degree of SMS, although the coping mechanisms learned by experienced astronauts do seem to help them feel better sooner (Davis et al. 1988; Reschke et al. 1998). No gender effects have been noted (Oman 1998). Fitness (measured by VO2 max) is not correlated with SMS susceptibility (Jennings et al. 1988), although there is a link with terrestrial motion sickness (Banta et al. 1987). Some retrospective studies have attempted to identify factors unique to astronauts who experienced SMS. Astronauts with a higher degree of ocular torsion asymmetry (larger differences in the rotation of the eye in its socket) were more likely to experience SMS. This finding suggests that the reflex controlling eye rotation (which comes from the otoliths) may be involved in SMS (Diamond and Markham 1991). Individuals who rely more heavily on visual cues of orientation (as opposed to kinesthetic or proprioceptive) feel more motion sickness (Harm et al. 1998).

The triggers of SMS have been difficult to determine, probably because of many conflicting reports in the literature. Astronauts on the early Mercury and Gemini flights did not report SMS symptoms (Clement 2003). It has been suggested that these astronauts may have underreported their symptoms (Clement 2003), but there were notable differences between their missions and more recent ones that might explain less illness. Probably the most notable difference is that the early astronauts experienced much higher G forces at landing and takeoff than recent crews do (~8 G as opposed to 3 G on the shuttle) (Clement 2003), but this does not seem likely to explain fewer SMS symptoms. In fact, ground-based human centrifuge studies have shown that high G exposure causes a suite of symptoms very similar to SMS (Bles et al. 1997).

More to the point, long-duration crewmembers experience more SMS with more severe symptoms than crewmembers on short missions (Ortega and Harm 2008), which suggests that duration in microgravity is an important causal factor. Early crews were also in an extremely small cabin that prohibited movement (Clement 2003). Later crewmembers have reported that SMS symptoms seemed to be initiated by vertical head movements (as opposed to head rotations) (Oman 1990; Bos et al. 2002), and it has been suggested that the Mercury and Gemini astronauts did not experience SMS symptoms because they could not move their heads in their confined vessel. It is now thought that head movements play a significant role in initiating SMS. It has been demonstrated that head movements in the pitch direction are more provocative of motion sickness than movements in the yaw or roll direction (Lackner and Graybiel 1986), and astronauts are advised to limit head movements, particularly those in the pitch direction (Clement 2003). In contrast, being confined (strapped to an upright tilt table) makes people feel much sicker when they are exposed to movement and moving visual fields in a ground analog of motion sickness (Faugloire et al. 2007). Taken together, these data leave us uncertain about the relative roles of microgravity, high G, head movement, and confinement in SMS.

Spaceflight Evidence

Since the mechanism of SMS is not well understood, it is hard to say which flight-related physiological changes are involved in SMS. Nevertheless, several physiological effects of spaceflight are likely related to SMS. First, and probably most important, during flight the otoliths of the inner ear no longer rest on the vestibular hair cells the way they do when in the pull of Earth's gravity. This repositioning is similar in many ways to what occurs in terrestrial benign paroxysmal positional vertigo, in which otoconia are dislodged from their normal position and shift into a semicircular canal, resulting in altered sensation and signaling from the vestibular organ. The symptoms are dizziness, spinning or moving sensations, vertigo, light-headedness, unsteadiness, loss of balance, blurred vision, nausea, and sometimes vomiting. They are often triggered by head movements, much like SMS (Parnes et al. 2003).

The best terrestrial treatment for benign paroxysmal positional vertigo is a series of movements that result in repositioning the misplaced crystals (Herdman 1990), but unfortunately this is not feasible in spaceflight. In addition to the gross physical change of otolith displacement in spaceflight, modern molecular techniques have shown that changes in gene expression occur during spaceflight. The number of cells in vestibular neurons that express Fos and other intermediate early gene products increased in rats that experienced 14 d of spaceflight. A corresponding decrease in expression occurred in brainstem reticular structures. These data indicate plasticity in response to the changed environment and also suggest that less proprioceptive input is available in flight (Balaban et al. 2002; Pompeiano et al. 2002).

Studies in which animals were reared in spaceflight have also shown evidence of plasticity in the otolith system (Wiederhold et al. 2003). The maculae also adapt to G forces and change synaptic connectivity (Ross and Tomko 1998). Exposure to hypergravity (2 G) for 2 h increased the amount of mRNA for H1 histamine receptors in rat hypothalamus. This is an intriguing finding, given that antihistamines are used therapeutically for SMS. Changes at the protein level were not examined in this study, and although the mRNA was normalized to glyceraldehyde-3-phosphate dehydrogenase (GAPDH), the authors did not show whether GAPDH mRNA was altered by the hypergravity exposure. They did show, however, that rats given a labyrinthectomy do not show this histamine receptor mRNA change (Sato et al. 2009).

Spaceflight Analog Studies

Most motion sickness testing has been conducted on the ground with volunteers, and because of this, is likely to have encouraged self-selection among volunteers (Hoyt et al. 2009). Most individuals have been in situations that might provoke motion sickness symptoms at some point in their lives. The people who suffer the worst symptoms would seem to be the least likely to volunteer for studies that are likely to make them feel ill. So it is possible that data collected from these studies is skewed in some fashion, but it is not known how or to what degree these data may be affected.

N&V syndromes are all clearly multisystem and thus do not lend themselves to reductionist models like tissue-culture or single-cell experiments do, especially when so little is known about the molecular mechanisms. A variety of animal models have been useful in N&V studies, but only a few of them involve motion-induced N&V; most involve a chemical stimulus. However, various measurements can be made during the brief free fall achieved in the dropped-animal model (Song et al. 2002; Anken and Hilbig 2004). A modified, scaled-down Ferris wheel device has been used in a number of studies of motion-induced N&V in cats (Crampton and Lucot 1985). Fish may have floats attached to their bodies to remove some of gravity's effects (Hoffman et al. 1980). Unfortunately, the data are scattered; either significant species differences occur in responses to motor stimuli or these stimuli fail to accurately reproduce the spaceflight experience. Given all of the variability, none of these models seem particularly useful for the testing of potential therapies for motion-induced sickness.

Several human models of motion-induced illness exist. Flight simulators (a moving platform coupled with a visual display) can realistically mimic many sensory experiences from real flight situations (except that the otolith input is not really changed). The rotating optokinetic drum is a model in which a person sits still inside a rotating drum with stripes painted on it. The mismatch of visual information and proprioceptive information is similar to the one in the proposed mismatch theory (the eyes see movement, but the body does not perceive movement) and motion sickness is produced (Hu et al. 1991). Nausea and motion sickness can be induced in a rotating drum with striped visual field (Stern et al. 1985; Hu et al. 1991). The most useful model seems to be cross Coriolis coupling, in which a person is seated in a rotating chair and asked to perform a series of head movements (Fernandez and Lindsay 1964).

Parabolic flight is used as a motion sickness stimulus in many studies, but because the percentage of the total time that is spent at high G is relatively long compared to the total time in microgravity, it is not a straightforward model of microgravity exposure. Parabolic flight has been shown to increase the stress hormones cortisol, prolactin, and ACTH (Schneider et al. 2007), which probably increase in spaceflight, but there has been no direct comparison. Regardless, the time periods in microgravity are too brief for this model to be useful for pharmacology experiments.

Earth-bound travel is sometimes used as a motion sickness stimulus, but so many of the population are accustomed to automobile and train rides that these stimuli are not very reliable in producing motion sickness symptoms. More individuals experience symptoms on boat rides, but since it is impossible to control the stimulus on a boat, large numbers of participants on numerous journeys would be required for a study to be considered conclusive.

Motion Sickness Treatments

Terrestrial motion sickness and nausea treatment have guided attempts at treating SMS. Tremendous improvements have been made in recent years in the treatment of nausea caused by cancer chemotherapy. The $5HT_3$ antagonists, ondansetron in particular, effectively reduce the chemically induced activation of the chemoreceptive trigger zone (CTZ) and/or act directly on the $5HT_3$ receptors of the enteric nervous system to reduce activity of the gut itself. Unfortunately, ondansetron does not seem to be effective in the treatment of nausea induced by motion (Stott et al. 1989; Levine et al. 2000; Reid et al. 2000; Muth 2006; Hershkovitz et al. 2009). In fact, a $5HT_3$ receptor antagonist was indistinguishable from a placebo in prevention of motion-induced nausea (Stott et al. 1989).

Agents that target the vestibular inputs to the vomiting center via muscarinic and histamine receptors have proven to be more efficacious in alleviating motion-induced symptoms (Reid et al. 2000). These include the antihistamines (promethazine) and the anticholinergics (scopolamine) that have been found effective for SMS (Davis et al. 1993a; Putcha et al. 1999).

ANTIHISTAMINES. Histamine antagonists (via the H1 receptor subtype) are effective for reducing motion-induced illness, but several second-generation non-sedating antihistamines from the 1990s were removed from the market because of QT interval prolongation and ventricular arrhythmia, probably via interaction with a potassium channel (Kohl et al. 1991). These risky antihistamines include terfenadine (Seldane®) and astemizole (Hismanal®).

Of the less-sedating current-generation antihistamines, chlorpheniramine significantly lengthens the time tolerated in a rotating chair, with side effects of dry mouth
and sleepiness. Chlorpheniramine delays reaction time on behavioral performance
tests, but no other performance impairments have been noted (Buckey et al. 2004).

Promethazine has activity at D2 dopamine receptors, H1 histamine receptors,
muscarinic acetylcholine receptors (Connolly et al. 1992), and inward rectifier
potassium channels (Jo et al. 2009), and there is some evidence for activity at the
benzodiazepine site of $GABA_A$ receptors (Plant and MacLeod 1994). Promethazine
(PMZ) does not affect autonomic functions such as arterial pressure, carotid baroreflex, and catecholamine concentrations (Brown and Eckberg 1997). It is perceived
as effective for prevention and treatment of SMS by flight surgeons and crewmembers (Davis et al. 1993a; Putcha et al. 1999).

Metoclopramide is a D2 antagonist and $5\text{-}HT_{3/4}$ antagonist, but it has proven ineffective on motion-induced nausea, at least in the rotating chair model (Kohl 1987).
Taken together, the motion sickness efficacy of various antihistamines suggest that
properties required for relief of motion sickness include both H1 antagonism and
D2 dopamine antagonism.

Antihistamines in general are associated with sedation, reduced reaction times,
and other performance impairments (Parrott and Wesnes 1987; Hindmarch and
Johnson 2001a; Ridout and Hindmarch 2003). In the rotating chair model of motion
sickness, operationally used doses of PMZ improved N&V symptoms but caused
impairments on a variety of behavioral tests involving cognition and reaction time
(Fig. 7.3). Impairments were similar to those caused by moderate doses of alcohol
(Cowings et al. 2000). Interestingly, stimulation of H1 receptors has been shown to
increase their expression level (Kitamura et al. 2004). It is not yet known if chronic
antihistamine use reduces expression level, but this is a question that should be
examined, especially if antihistamine use is of long duration.

ANTICHOLINERGICS. Scopolamine, a muscarinic antagonist, is probably the most
common terrestrial treatment for motion sickness, especially in the transdermal
patch form used prophylactically by many travelers prone to motion sickness.
Sedation is frequent side effect of scopolamine, and delayed reaction times have
been measured (Parrott and Wesnes 1987; Howland et al. 2008). Diphenhydramine
and the related compound dimenhydrinate are more popular for oral relief (Katzung
2007). Buccal scopolamine has been shown effective in parabolic flight (Norfleet
et al. 1992). All of these are thought to act via muscarinic receptors on vestibular
neurons. Interestingly, administration of physostigmine, an acetylcholinesterase
inhibitor that prolongs the activation of cholinergic receptors, produces an SMS-
like syndrome (Janowsky et al. 1984). Muscarinic receptors are also involved in
peristalsis, the muscular contractions of the GI tract that propel the contents through
the system (Crema et al. 1970). This is probably the mechanism of the constipation
and urinary-retention side effects associated with these anticholinergic motion
sickness treatments.

SYMPATHOMIMETICS. Sympathomimetics are drugs that mimic the effects of epi-
nephrine, norepinephrine, or dopamine, the neurotransmitters of the sympathetic
nervous system (Katzung 2007).

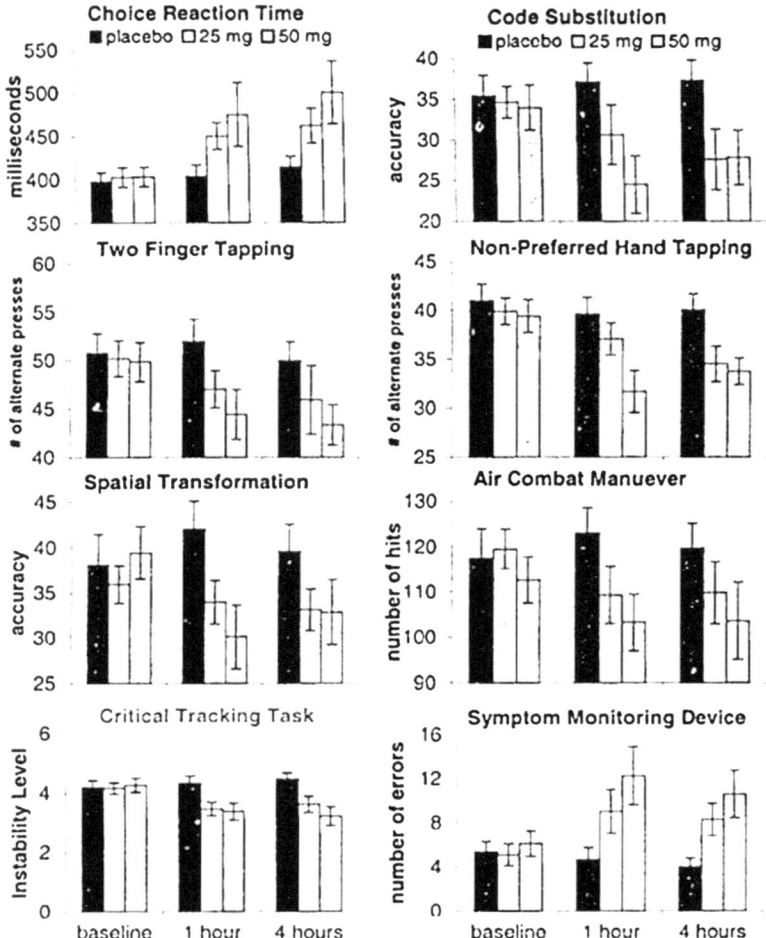

Fig. 7.3 Results of behavioral tests with placebo, 25 or 50 mg promethazine (Cowings et al. 2000). Used with permission, Aerospace Medical Association

Methamphetamine and phenmetrazine, both stimulants that increase release of norepinephrine and dopamine, have been shown to increase tolerance to motion sickness-inducing stimuli in ground studies, whereas related compounds methylphenidate (Ritalin) and phentermine (part of Phen-Fen) did not (Kohl et al. 1986). Although the motion sickness stimuli and measurements in this study were all extremely well designed, the pharmacology was "single-point"; that is, only a single dosage was used for testing each drug, and there was evidently no attempt to match effective doses of the various drugs used (Kohl et al. 1986). It is possible that methamphetamine was effective in this study because it was used at a relatively high dose, and that the agents found ineffective were given at doses that were too low.

The use of single-point pharmacology makes it impossible to draw a conclusion from this study.

Sympathomimetics in combination with antihistamine or anticholinergic seem to be more effective than either agent alone (Wood and Graybiel 1968; Graybiel et al. 1975). Performance errors associated with PMZ use were diminished by concomitant administration of amphetamine (Wood et al. 1984; Schroeder et al. 1985). Modafinil plus scopolamine allowed 29% more motion tolerance in a rotating chair (Hoyt et al. 2009). Modafinil is a treatment for narcolepsy and attention deficit hyperactivity disorder, and has functional similarities to amphetamine in that it increases release of dopamine and norepinephrine, but it also releases histamine.

THE RENIN-ANGIOTENSIN SYSTEM. Components of the renin-angiotensin system seem to have effects on motion sickness, but it is not clear whether they act independently or via the sympathetic nervous system. Treatment with angiotensin II increases arterial blood pressure by constricting peripheral blood vessels via a sympathetic mechanism and/or vasopressin release (Katzung 2007). This could explain the actions of the sympathomimetic drugs listed above, and it provides new possible sites for therapeutic intervention.

Atrial naturietic peptide (ANP) is secreted by the heart in response to sympathetic drive, high sodium, or atrial distension (Guyton and Hall 2006). ANP has been shown to increase with time spent in head-down-tilt bed rest, on a scale of minutes (Grundy et al. 1991), probably because of atrial stretching caused by fluid redistribution. A high-sodium diet suppresses sympathetic nervous system activity and the renin angiotensin system. This raises the possibility that a high-salt diet could prevent the increase in sympathetic output that is required to compensate for postflight orthostatic hypotension. However, ANP treatment reduces capillary filtration and keeps more fluid in the capillaries, which would tend to prevent orthostatic problems (Watenpaugh et al. 1995). Arginine vasopressin (AVP), a hormone released by the brain in response to dehydration, causes nausea and lethargy (Kohl et al. 1991). AVP antagonists have been tested as motion sickness treatments, but current forms are expensive and difficult to deliver to the brain (Kohl et al. 1991).

NEUROKININ ANTAGONISTS. The neurokinin-1 receptor is found in the brainstem and in the GI tract, both areas involved in N&V. Neurokinin antagonists were shown in the 1990s to be effective in animal models of motion sickness (Gardner et al. 1996; Lucot et al. 1997). Maropitant prevents vomiting in dogs on car rides (Conder et al. 2008) and in dogs exposed to chemicals (equivalent to ondansetron) (Sedlacek et al. 2008). These drugs are starting to be used clinically with cancer chemotherapies, especially with 5HT$_3$ antagonists. However, in human trials of motion-induced nausea, neurokinin antagonists GR205171 and L-758, 298 were no different from a placebo (Reid et al. 2000). Neurokinin-1 antagonist GR205171, alone or combined with ondansetron, was not effective for chair rotation motion-induced nausea (Reid et al. 2000).

OTHER MOTION SICKNESS TREATMENTS. As might be expected for a disorder that has affected so many people for centuries, many remedies have been tried, and some have showed some success. GABA receptors are found in the brainstem, and the GABA agonist and epilepsy drug phenytoin has been used by NASA as a seasickness remedy (Woodard et al. 1993). Taken together with the recently proposed PMZ activity at the benzodiazepine site of GABA receptors (Plant and MacLeod 1994), it seems that GABA receptors may play a role in motion-induced illness. Cannabinoid receptors are found in the brainstem and cannabinoid antagonists have been successfully used as N&V treatments, but have not yet been tested for motion-induced illness (Heer and Paloski 2006, 2006). The GABAB receptor agonist baclofen has also been shown to reduce motion sickness occurrence, and its mechanism of action may overlap with that of the cannabinoid agonists (Cohen et al. 2008). Stugeron is an internationally used over-the-counter motion sickness remedy. It is an antihistamine, a calcium channel antagonist (L-type), a D2 dopamine antagonist, and a muscarinic antagonist (note the overlap with PMZ actions). Vinpocetine (cavinton) is in the family of nootropic drugs and shows some effectiveness in ground models of motion-induced illness (Matsnev and Bodo 1984). Esiniferatoxin, a plant-derived capsaicin analog, is antiemetic in motion sickness in a shrew model (Andrews et al. 2000).

However, all of these potential nausea therapies and preventatives carry the likelihood of significant side effects, including performance-impairing sleepiness (Gilman et al. 1990; Katzung 2007) as well as reductions in speed and accuracy of performance (Paule et al. 2004). Because of these potential adverse effects, the mission pilot, commander, and flight engineer are not permitted to take antimotion sickness medications before launch (Ortega and Harm 2008).

SPACE MOTION SICKNESS TREATMENTS. Scopolamine, a muscarinic antagonist, and the antihistamines diphenhydramine and dimenhydrinate have been used during flight and found to be somewhat effective. The sedating side effects of scopolamine were problematic until it was reformulated with dextroamphetamine, previously used as a stimulant in flight medicine. Scopolamine/dextroamphetamine (scop/dex) has not been widely used since 1993 (Davis et al. 1993a) because it does not work for everyone (ineffective for 38%; Nachum et al. 2006) and withdrawal from it causes unpleasant rebound-type side effects (Davis et al. 1993a).

Promethazine (Phenergan®) is the antiemetic most used in spaceflight. Several delivery forms have been made available to astronauts over the years, but currently, most doses taken are intramuscular for immediate symptom relief. Some astronauts use it prophylactically in the first few days of flight, particularly before sleep. Few complaints of sedation have been associated with its use in space (Davis et al. 1993b); this finding was considered surprising by flight surgeons and was presumed to be due to the excitement of flight (Bagian and Ward 1994). This explanation has not yet been confirmed; it remains possible that in space this drug is handled differently by the body and absorbed, distributed, metabolized, or excreted in a fashion that reduces the typical sedative side effect.

Promethazine has been used as a motion sickness remedy for decades; in 1968 it was recognized as the best remedy available not only for passengers but also for crewmembers on the Cunard shipping line (Heggie and Entwistle 1968). Promethazine has activity at D2 dopamine receptors H1 histamine receptors, and there is some evidence for activity at the benzodiazepine site of $GABA_A$ receptors (Plant and MacLeod 1994).

Typically, when the mechanism of an ailment is unknown but a useful treatment is available, the characteristics of the treatment can be used to help understand the original problem. Unfortunately, this wide range of PMZ activities does not help to elucidate the mechanism of motion-induced nausea or to minimize untoward effects. In fact, a recent investigation into the causes of postflight orthostatic hypotension and syncope shows that PMZ increases the likelihood of syncope (Shi et al. 2010). Furthermore, since PMZ has activity at $GABA_A$ receptors and is sometimes taken with sleep aids that act on the same receptors, there are possibilities of potentially dangerous drug interactions.

Problems with Current SMS Therapies

Scopolamine does not work for everyone; in fact it is largely ineffective for about one-third of the people who try it. A rebound effect has been noted upon termination of scopolamine/dextroamphetamine, a brief intensified return of the original symptoms (Wood et al. 1986). It is currently not suggested by flight surgeons as first-line treatment and has not been used much since 1986 (Davis et al. 1993a). Dry mouth, drowsiness, and blurred vision are fairly frequent complaints with scopolamine, even with transdermal delivery intended to prevent concentration peaks seen with other modes of administration (Clissold and Heel 1985).

Promethazine side effects are numerous and, in terrestrial use, experienced by most people. The most frequently reported side effects include sedation, confusion, blurred vision, dry mouth, faintness, urinary retention, tachycardia, and bradycardia. Statistically significant decrements in performance compared to the placebo were observed on 10 of 12 tasks when subjects were given 25 or 50 mg of PMZ (Cowings et al. 1996). Performance decrements were equivalent to those of individuals with blood alcohol levels of 0.085% for 25-mg and 0.13% for 50-mg dosages (Cowings et al. 1996) (both levels are above "driving under the influence" limits in most states). Also, terrestrial PMZ use has been linked to urinary retention, which has occurred during flight, as well as other urinary problems (Stepaniak et al. 2007).

Furthermore, a recent investigation into the causes of postflight orthostatic hypotension and syncope shows that PMZ decreases the likelihood of finishing a 20-min tilt test (five of eight control subjects completed the tilt test, but none of eight subjects given PMZ were able to complete the test) (Shi et al. 2010). Given that some crewmembers require assistance when exiting the vehicle after an uneventful landing, there is a concern that orthostatic intolerance could become a serious impediment in the case of an emergency egress. It is clear that our current treatment strategies for SMS are less than ideal. However, until new drugs are shown to be efficacious for motion-induced illness, the side effects of PMZ must be ameliorated or tolerated as well as possible.

Gastrointestinal Summary

The evidence regarding SMS occurrence and treatment efficacy falls under Category II (from controlled studies). Much of the evidenced presented above regarding terrestrial motion sickness treatments fall under Category I (derived from randomized, controlled trials).

Space motion sickness is one of the chief complaints of crewmembers, particularly in the gravitation transition periods of after launch and after landing. It and the side effects of the medications used, limit crew activities during these times. Ideally, a new and better therapy should be identified. Until that time, a sensible strategy for alleviating promethazine side effects should be established. The root cause of SMS is still not clear. As in all terrestrial medicine, an efficacious and specific treatment is much more likely to be developed if the mechanism of the disorder is understood and the molecular players have been identified.

Although PMZ is mostly effective for SMS treatment while crewmembers are on board, it bears the risk of causing potentially life-threatening side effects in the case of emergency egress. This knowledge gap will remain a concern until there is a treatment that is both effective and safe. The possible interactions of SMS treatments with other drugs administered on board are not yet known. This is especially important in the case of sleep medications, which mostly act via GABA receptors, also a target of PMZ. Antihistamines given for head congestion are an additional concern, since PMZ also targets histamine receptors. The extent to which space motion sickness medications play a role in inflight urinary problems is not yet known. Given the known terrestrial side effects of these drugs, they should be investigated as potential causes of or contributors to urinary retention and other urinary disorders.

Chapter 8
Musculoskeletal System

Skeletal System

Briefly, the adult human skeleton undergoes constant remodeling in response to physiological conditions, signaled by hormones, plasma ion concentrations, and physical stresses. The Bone Discipline has already prepared Risk Reports on the challenges faced by the skeletal system in spaceflight, and how its physiology might be altered by spaceflight (LeBlanc et al. 2007; Sibonga 2008a, b). Astronauts, particularly those on long-duration missions, have suffered significant bone loss in the past. Understanding the triggers for this and developing appropriate counter-measures has been a concern for decades.

Two very different spaceflight analogs have proven to be good models of space-flight conditions for the musculoskeletal system and have been heavily used: the hind limb-unloaded rodent model, and the human head-down-tilt bed rest model (LeBlanc et al. 2007; Spector et al. 2009). Bone-loss mechanisms and countermea-sures involving sodium, calcium, and vitamins D and K have been addressed by the Nutritional Biochemistry Discipline (Smith and Zwart 2008), and others that involve load-bearing exercise were described by the Muscle and Exercise Discipline (Discipline 2008). Advances are being made quickly in this field. Recently, oxyto-cin, thought to be involved only in lactation and social behavior, has been shown to directly participate in regulation of bone density (Tamma et al. 2009), but much remains to be discovered about its role and how it may be manipulated pharmaco-logically. Only the special topics related to use of pharmaceuticals for bone preser-vation will be addressed here.

Spaceflight Evidence

TERIPARATIDE. Human parathyroid hormone (PTH) is an important regulator of cal-cium and phosphate in bone. It has different actions depending on duration and con-centration: chronic low concentrations increase both intestinal absorption of dietary

V.E. Wotring, *Space Pharmacology*, SpringerBriefs in Space Development,
DOI 10.1007/978-1-4614-3396-5_8, © Virginia E. Wotring 2012

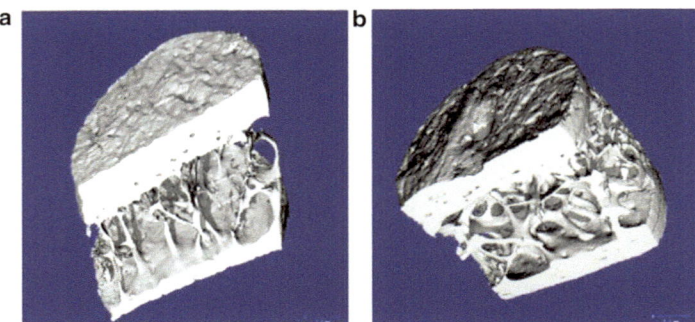

Fig. 8.1 3D computed tomography reconstructions of a paired iliac crest biopsy from a 65-year-old woman, (**a**) before treatment, and (**b**) after 21 months of teriparatide treatment. Note the change in connectivity and internal structure (Jiang et al. 2003). Used with permission, Wiley

calcium and renal reabsorption of urinary calcium, whereas bolus administration will stimulate osteoclasts to initiate more bone turnover (Guyton and Hall 2006). These apparently contradictory effects are likely due to molecular regulatory mechanisms that are currently unknown. Advances are being made rapidly in this area; dozens of new molecules that participate in these mechanisms are being discovered.

In clinical trials with postmenopausal women, teriparatide has been shown to reduce vertebral fractures by 65% (Neer et al. 2001) and increase bone mineral density (BMD) by ~14% in the lumbar spine and ~6% in the femoral neck (Jiang et al. 2003). PTH and its analogs are used as osteoporosis therapies. Teriparatide is a recombinant form of just the first 34 amino acids of PTH and has been shown to increase the population of osteoblasts (bone-building cells) by stimulating their replication and at the same time inhibiting osteoblast apoptosis (programmed cell death). This strengthens the bone structure (Jiang et al. 2003), rather than just increasing the BMD as the bisphosphonates do (Fig. 8.1). This outcome is obviously desirable, but these drugs are still in relatively early stages of development. Little is known about risks associated with using them for more than 24 months, or with using them in combination with other bone-strengthening strategies (Brixen et al. 2004). Also, these early human trials have been conducted mostly in postmenopausal women; little information is available regarding the effects of PTH analogs on younger healthy people.

Additionally, because PTH and its analogs are peptides, they are currently available only in injectable form and must be refrigerated, a significant operational problem for flight use. The most common side effects noted with teriparatide use are a mild and transient hypercalcemia and leg cramps. Reports exist of increased risk of osteosarcoma in rats, but this has not been supported by human data. Furthermore, a side effect of orthostatic hypotension has been noted in at least one study, but it seems to occur only during the first few hours after dosing (Lilly 2004). Scheduling a dose-free period near landing would likely avoid this problem, but this scenario would require testing.

BISPHOSPHONATES. These have now been used clinically for more than a decade. Alendronate was the first bisphosphonate to come into common use in the treatment of osteoporosis associated with age, and it is associated with increased BMD in older people (Liberman et al. 1995; Heaney et al. 1997). Some studies also indicate a reduction in fractures (Cummings et al. 1998; Black et al. 2000), although this is under current investigation by the FDA (as of 3/2010). Human head-down-tilt bed rest studies with younger subjects have also shown a reduction of the disuse-induced loss of BMD seen in controls and a reduction in urinary calcium, as well as improvement in other measurable bone markers (LeBlanc et al. 2002). Because of these results, alendronate is currently being used in a flight study. An additional benefit of this kind of treatment and its reduction in urinary calcium excretion (LeBlanc et al. 2002) is a reduction in risk of formation of renal stones (Senzaki et al. 2004). Pamidronate has been shown to similarly reduce urinary calcium and renal stone formation, and to improve BMD in a 90-d head-down-tilt (HDT) bed rest study (Watanabe et al. 2004). The combination of dietary countermeasures, exercise, and bisphosphonates in various combinations are also being tested (Rittweger et al. 2005). The effect of long-term bisphosphonate therapy in younger people is not yet known. There is some evidence of increased risk of jaw necrosis in dental patients (Fusco et al. 2009; Treister et al. 2009), and increased risk of long bone fracture in older osteoporotic patients (Odvina et al.), but it is not known if crewmembers would experience any increase in their risk of either of these.

New-Generation Bone-Preserving Drugs: Anti-RANKLs

Inhibitors of RANK (Receptor Activator for Nuclear Factor κ B Ligand) are a new type of bone-preserving therapy. RANK is a molecular signal produced by osteoblasts that is an activator of osteoclasts (bone-resorbing cells) (Bai et al. 2008). The anti-RANK or anti-RANKL (RANK ligand) agents under development bind to the RANK molecule and inhibit its usual function, activation of osteoclasts (Pageau 2009), thus decreasing bone resorption. The most well developed of these is a monoclonal antibody called denosumab. Denosumab has been shown to increase BMD, but safety trials have not yet been completed (McClung et al. 2006; Pageau 2009). RANK signaling also plays a role in immune system function, so careful attention should be paid to possible immune system effects. Also, there has been a report of jaw necrosis with denosumab, similar to what has been reported during use of bisphosphonates (Taylor et al. 2009). Safety trials and statistics from clinical use are especially important for the new RANKL drugs.

Reduction of Bone Turnover and Reduced Renal Stone Risk

An increased incidence of renal stones has been noted, both during missions and after return from spaceflight (Pietrzyk et al. 2007). It is thought that these stones are caused by unusually high urinary calcium concentrations related to high bone turnover rates. Potassium citrate therapy has been used terrestrially to complex urinary calcium into a soluble form that is excreted in the urine, eliminating precipitation of crystals in the body (Sellmeyer et al. 2002; Zerwekh et al. 2007). It has also been used in long-duration crewmembers and shown to decrease urinary calcium, but the direct relationship with reduction in renal stones has yet to be established (Whitson et al. 2009). Bisphosphonates have been shown to reduce urinary calcium in terrestrial studies (LeBlanc et al. 2002; Watanabe et al. 2004), but it remains to be demonstrated whether the same holds true for teriparatide and denosumab. Whether a reduction in urinary calcium also occurs with spaceflight use of teriparatide and denosumab is not yet known, but this is one factor being tested in the current flight bisphosphonate study. It would be assumed that reduction of urinary calcium would reduce the risk of renal stone formation during and after spaceflight, but this also has yet to be confirmed.

Muscular System

Spaceflight Evidence

The Muscle Discipline has already prepared a Risk Report on the changes in muscle seen in spaceflight (Discipline 2008). Muscle atrophy has been associated with spaceflight since the early days of exploration and was initially thought to be the result of inactivity in a confined vehicle coupled with gravitational unloading of the muscles (Discipline 2008).

Muscle atrophy reduces physical strength, but when the diaphragm is affected, it also reduces the amount of oxygen that may be brought into the body. Similarly, when the heart muscle is affected, muscle atrophy reduces the rate and amount of blood flow throughout the body (Guyton and Hall 2006). Muscle atrophy is perceived as a significant threat to missions not only because it can limit the physical work a crewmember could perform, especially during EVAs and vehicle egress, but also because the reduction in cardiopulmonary reserve capacity could limit the body's ability to fight infection or heal injury (Discipline 2008; Gopalakrishnan et al. 2010).

The mechanism of spaceflight's effect on muscle physiology has not yet been determined. The same spaceflight analogs used in bone research, the hind limb-unloaded rodent model and the human head-down-tilt bed rest model, have been heavily used to study muscle disuse atrophy (LeBlanc et al. 2007), but neither of these reproduces all of the effects seen in spaceflight. The wide array of fluid, enzyme, and hormone shifts seen in spaceflight include many molecules that can influence muscle maintenance and growth (Leach 1981; Leach et al. 1991a; Chi et al. 1992). An increase in intracellular calcium has been measured in hind limb suspended gerbils, but it is not known if this occurs during spaceflight (Ogneva et al. 2010).

Any of the hormones, enzymes, or ion regulatory molecules are potential targets for interventions and countermeasures. Function of a calcium-activated potassium channel in a mouse is altered by hind limb suspension (Tricarico et al. 2005), but whether this occurs in spaceflight is unknown. Although pharmacological intervention intended to preserve muscle mass and strength is a relatively new area of research, some pharmaceuticals are under study. These include testosterone and its derivatives (Madeddu and Mantovani 2009). Safety and efficacy trials are being conducted with these drugs in patients with various wasting disorders and diseases, a group very different from space crewmembers. Any or all of these drugs may nevertheless prove useful in very fit crewmembers experiencing muscle atrophy associated with long-duration spaceflight. However, given the significant hepatic and renal side effects noted with the use of these kinds of drugs in the past, particularly when they are given by oral administration, systemic safety must be demonstrated in FDA and other ground trials before NASA experiments can begin.

Testosterone as an Anabolic Steroid

It has long been known that testosterone encourages the growth of larger muscles (Guyton and Hall 2006), and it has been used as a supplement in cases of muscle atrophy caused by cancer, other wasting diseases, chronic steroid therapy, and even aging (Bhasin et al. 1997; Hajjar et al. 1997; Sih et al. 1997; Snyder et al. 2000; Wang et al. 2000; Casaburi et al. 2004; Wang et al. 2004; Bhasin et al. 2006).

One problem with testosterone therapy for muscle growth is that muscle is not the only tissue affected. Testosterone affects the testes, prostate, skin, hair, bone, muscle, and the brain (Chen et al. 2005), and a great concern in treating older men, in particular, with testosterone is that the risk of prostate hypertrophy or cancer could be increased. Testosterone production has been shown to decrease in animals after relatively short flights (Macho et al. 2001), in animals in the hind limb suspension flight analog (Wimalawansa and Wimalawansa 1999), in the rotating tissue culture model (Ricci et al. 2004; Ricci et al. 2008), and it is also known to decrease in human males after about age 30, at a rate of ~1% per year (Morley et al. 1997; Roy et al. 2002).

Testosterone treatment seems to affect older men (with declining natural testosterone production) and younger men similarly (Bhasin et al. 1996; Bhasin et al. 2005). Although serious side effects (serum lipid derangements, "steroid rage") have been reported when testosterone was used without medical supervision, these cases seem to have involved doses many-fold higher than the doses used in clinical trials to boost muscle mass and strength, and these side effects have not been noted with therapeutic doses in a clinical setting (Tricker et al. 1996). Some studies indicate that testosterone can be supplemented at moderate doses in healthy younger men without significant side effects (Bhasin et al. 1996; Bhasin et al. 2001). Testosterone was used in one bed rest study and shown to preserve muscle mass and nitrogen balance, but there was still a loss in muscle strength (Zachwieja et al. 1999). Recent research into the molecular players involved in androgen effects on muscle indicate that, because of tissue-specific promoter sequences, it may be possible to target muscle with relative specificity, thus avoiding potential problems with prostate or other tissues (Li et al. 2007; Hong et al. 2008).

New Anabolic Compounds: Selective Androgen Receptor Modulators

Recent development of a new nonsteroidal class of selective androgen receptor modulators (SARMs) has raised the possibility of therapy more specific to muscle tissue—a drug that would be anabolic but less associated with neural, hepatic, or renal side effects (Dalton et al. 1998; Edwards et al. 1998; van Oeveren et al. 2006; Li et al. 2007; Bhasin and Jasuja 2009). The first-generation drugs of this type are in various stages of research (Narayanan et al. 2008). A slightly modified form of testosterone, 19-Nor-4-androstene-3-β,17-β-diol has been shown to increase lean body mass and bone mineral density in orchidectomized rats, without affecting prostate mass (Page et al. 2008). Another SARM, LGD2226, has been shown to increase muscle mass, BMD, and bone strength in rats, with no significant effect on the prostate or sexual behavior (Miner et al. 2007).

Several first-generation SARMs are in Phase I trials, mostly in patients experiencing various wasting diseases. Nandrolone decanoate increased lean body mass in dialysis patients (Johansen et al. 1999), a population that typically experiences wasting. Early indications are that these SARMs do not yield as much increase in muscle mass as testosterone. It may be that these kinds of drugs will not prove useful until a second or third generation of compounds have been tested (Bhasin and Jasuja 2009).

Musculoskeletal System Summary

The evidence regarding use of pharmaceuticals to reduce bone loss is Category I, from randomized, controlled studies, but is ground-based. There is an ongoing flight study testing the use of bisphosphonates during long duration spaceflight. It is not yet known if bisphosphonates (or new-generation alternatives), or inflight potassium citrate treatment reduces the inflight and postflight risk of renal stones to an acceptable level. It is not yet known if bisphosphonates given during flight (and after flight?) will mitigate bone loss so that osteoporosis and fracture risk are reduced in astronauts. Testing the new bone-preserving drugs should be done as they come through the FDA approval system.

The evidence regarding muscle changes in spaceflight fall under Category II, derived from controlled studies. It is not yet known if testosterone mitigates muscle loss in spaceflight without causing significant side effects. Newer agents with a more specific mechanism of action may provide better protection and/or fewer side effects. Testing of new muscle-preserving drugs should be done as they come through the FDA approval system.

Chapter 9
Multiple Systems Spaceflight Effects

While none of the physiological systems discussed previously operates independently of the rest of the body, the immune system is remarkable for its integration with many other areas of physiology. Furthermore, one of the hazards associated with leaving low-Earth orbit is increased exposure to radiation, which has effects on multiple physiological systems, including the immune system.

Immune System

The Immunology and Microbiology Disciplines have already prepared Risk Reports on the challenges faced by the immune system in spaceflight, and how its function might be altered by spaceflight (Crucian 2009). Since immune system function is affected by many environmental and physiological factors (Fig. 9.1), this area is particularly complex. Furthermore, it has recently been shown that some microorganisms the immune system has to fight are also affected by spaceflight (Nickerson et al. 2000; Wilson et al. 2002; Nickerson et al. 2003). Many physiological parameters seem to be involved, but only the special topics related to pharmaceutical use will be addressed here.

Spaceflight Evidence

It is clear that long-duration spaceflight is associated with oxidative cell damage (Stein and Leskiw 2000), but the cause is yet to be determined. Spending time outside Earth's atmosphere increases exposure to radiation, which is discussed in the Radiation section below. However, it is also possible that microgravity itself plays a role in producing oxidative damage, and rotating cell culture has provided the evidence. This kind of culture system, called a clinostat or bioreactor, keeps suspended cell cultures in constant free fall, mimicking what happens to bodies in orbit.

V.E. Wotring, *Space Pharmacology*, SpringerBriefs in Space Development, 71
DOI 10.1007/978-1-4614-3396-5_9, © Virginia E. Wotring 2012

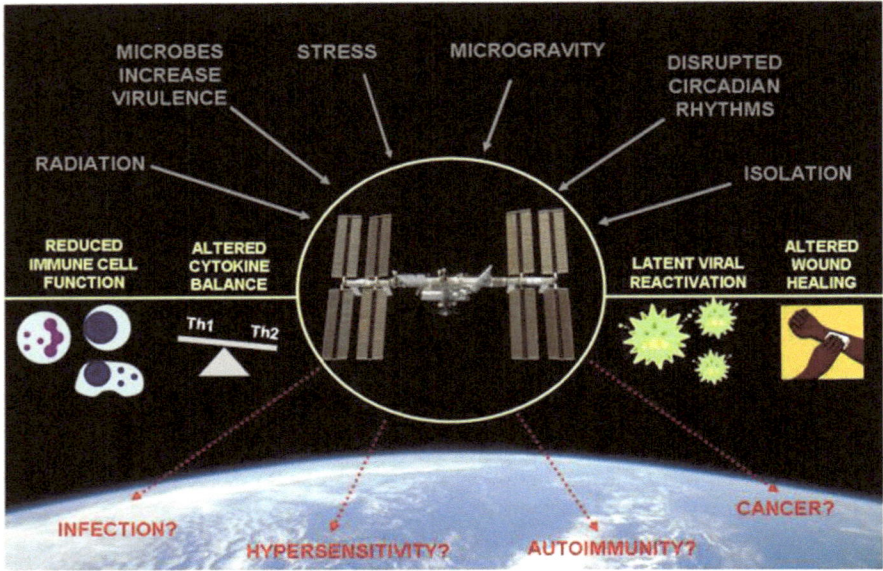

Fig. 9.1 Diagram of the many factors influencing function of the immune system in microgravity (Crucian 2009). NASA, Open Access

An increase in nitric oxide production in rotating cultures has been demonstrated in cardiac cells and brain cell model systems (Wang and Good 2001; Xiong et al. 2003).

Nitric oxide is a signaling messenger molecule but plays a role in pathology when free-radical formation becomes too great. Reactive nitrogen compounds participate in a number of stress pathways, usually beginning with a nitric oxide messenger that reacts to form a free radical that interacts with tyrosine residues of proteins. These nitrosylated proteins may exhibit altered function, presumably because of the change in their structure (Greenacre and Ischiropoulos 2001). These increases in oxidative products are very similar to those seen with increased exposure to radiation, and it is likely that the same therapies will prove useful, regardless of whether the root cause is microgravity or radiation.

Function of the immune system is altered by spaceflight, as described in the Immunology Discipline Risk Report (Crucian 2009). Stress hormone concentrations are elevated after flight (Fig. 9.2) (Pierson et al. 2005), especially long-duration flights. Natural killer cell, monocyte, and neutrophil function are all reduced after flights (Crucian 2009). The trigger is not yet understood, although candidates include radiation exposure and chronic stress in addition to microgravity itself. Many similar changes in redox state and immune cell function are seen after exposure to radiation, and the same kinds of pharmacological interventions are likely to be useful (see Radiation section for discussion).

It is clear that changes in immune system regulation occur in spaceflight that would tend, on Earth, to result in a higher rate of infection. It is not yet known if

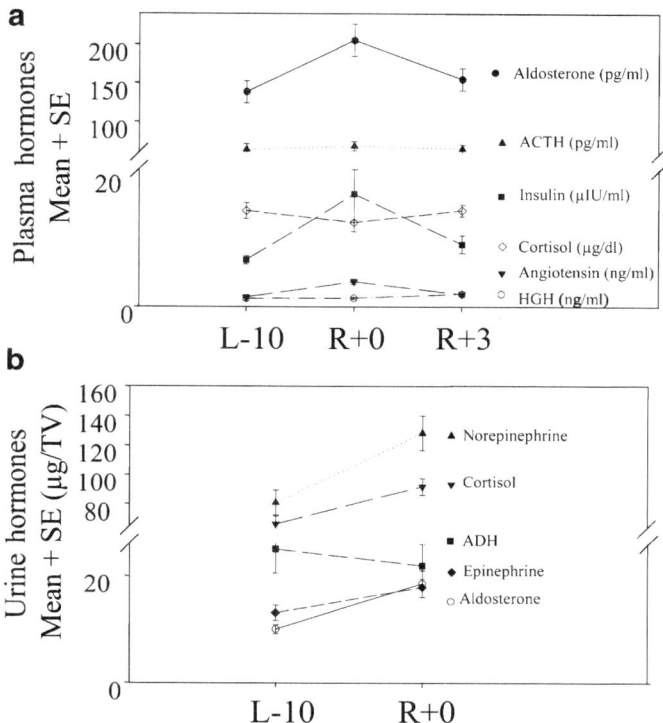

Fig. 9.2 Plasma (**a**) and urine (**b**) concentrations (mean ± SE) of stress hormones of 32 astronauts before and after space flights. TV, total volume (of 24 h pool) (Pierson et al. 2005). Used with permission, Elsevier

infection rates are higher in spaceflight, or if pharmaceutical treatments or preventatives should be considered, but this remains a possibility. Of particular concern is the reactivation of latent viruses demonstrated during and after spaceflight. Elevated copy numbers have been shown with the herpes viruses varicella-zoster virus, cytomegalovirus, and Epstein-Barr virus (Fig. 9.3) (Mehta et al. 2004; Pierson et al. 2005), although one study indicates that the increased viral shedding is experienced by more crewmembers in the period shortly before flights (Payne et al. 1999). It is not yet known if symptoms are likely after viral shedding and reactivation, or if transmission to others is more likely, so the need for therapeutics is currently unknown.

It has recently become clear that microbes themselves are altered by spaceflight (Crucian 2009), raising the possibility that the efficacy of the mission pharmaceuticals used against them might be changed. Nickerson and colleagues have clearly shown that *Salmonella* virulence changes in the low-shear rotating bioreactor system, with 100% death by day 11 in the animals injected with treated cultures compared to day 21 for control cultures (Nickerson et al. 2000). This system is accepted as a good spaceflight analog for liquid cell culture (Nickerson et al. 2003; Nickerson et al. 2004) and has permitted many experimental situations to be tested

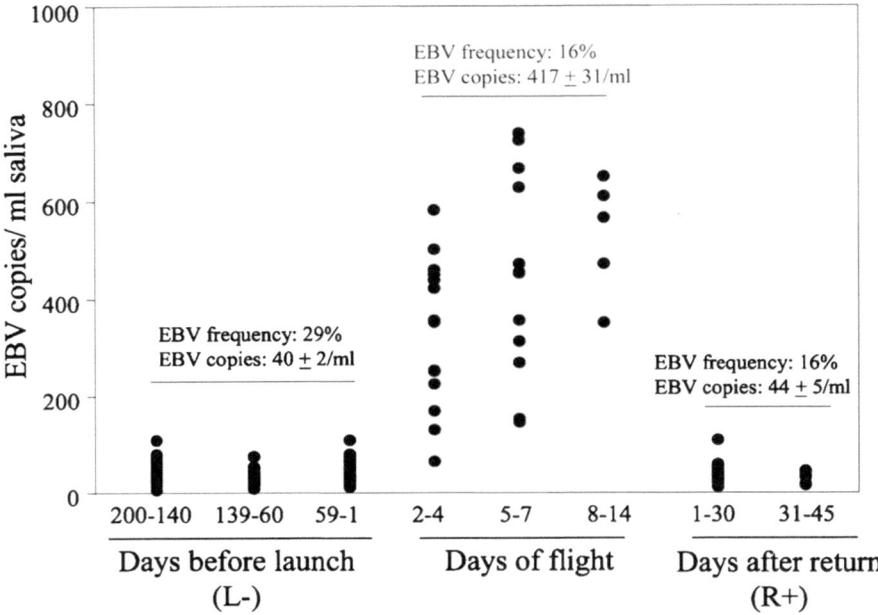

Fig. 9.3 Distribution of the number of EBV copies per mL of saliva in EBV-positive samples from 32 astronauts during sampling periods before, during, and after 10 Space Shuttle missions (Pierson et al. 2005). Used with permission, Elsevier

before flight. Nickerson and colleagues have followed their *Salmonella* virulence finding with identification of the genes and proteins involved using modern molecular biology techniques (Wilson et al. 2002; Nickerson et al. 2003). This work is still in the early phases, but the indications are that these changes also occur in spaceflight (Wilson et al. 2007) and expression of many genes and proteins is affected.

Genes involved in RNA transport and stability, iron utilization, cell motility, transmembrane transport, and redox homeostasis have been implicated, but what these changes mean for the function of cells is not yet known (Wilson et al. 2008). A chaperone of RNA molecules (and thus a regulator of RNA translation to protein)—Hfq—has been shown to decrease modestly in spaceflight cultures (Wilson et al. 2007) and is particularly intriguing because it has the potential to affect the eventual expression of many proteins. Further studies are required, not only to pinpoint the genes and proteins altered by spaceflight but also to determine the physiological role of these changes, to determine which aspects of spaceflight trigger the changes, and to examine additional species of microorganisms.

It should be stressed that these studies are in the early qualitative phases. Publication standards for appropriate quantification and normalization of gene expression and microarray results have recently been updated and strengthened and will require a somewhat more rigorous approach for future experiments in this area (Knudsen and Daston 2005; Bustin et al. 2009; Bustin 2010).

Do Antibiotics Work Against Microbes Altered by Spaceflight?

Antibiotic testing of spaceflight-altered bacteria has not yet been tested systematically, but ensuring that the appropriate measures for treating infections are carried aboard must certainly be considered a high priority. If expression of genes or proteins involved in antibiotic resistance or plasmid exchange were found to be upregulated by spaceflight, antibiotic testing would become an even higher priority. It has been shown that gentamicin efficacy was the same in spaceflight as on the ground, although the flight cultures appeared to experience additional growth (Kacena and Todd 2000). However, in cultures of *S. aureus* and E coli from a crewmember, higher concentrations of kanamycin, colistin, erythromycin, chloramphenicol and oxacillin were required. These authors of this study also noted faster growth of microbes in flight conditions (Tixador et al. 1985). These studies were all performed during spaceflight.

Another study that examined susceptibility of various microbes to antibiotics after a long duration spaceflight found that *S. aureus* was more resistant to several antibiotics, but the other organisms they tested were more susceptible (Juergensmeyer et al. 1999). It seems likely that the variety of experimental conditions is confounding interpretation of results in this area.

Antibiotic Effect on Native GI Flora

It is known from terrestrial clinical use that antibiotic therapy can change the GI flora, sometimes leaving the patient more vulnerable to attack by a different set of microorganisms (reviewed in Sullivan et al. 2001). It has even been suggested that the gut microflora are actively involved in immune system function in allergy (Noverr and Huffnagle 2004). Whether antibiotic-induced flora changes occur during spaceflight has not yet been directly tested. It is also unknown if the flight-induced changes in immune system or microbial gene expression might lead to additional problems with secondary infection or inappropriate re-colonization.

Multisystem Radiation Effects

Spaceflight Evidence

The Radiation Discipline has already prepared Risk Reports on the challenges posed by radiation exposure during spaceflight. They estimate the dosage from an average 2–3 year exploration mission would be 0.5–1 Gy (Wilson et al. 1995).

Briefly, it is clear that exposure to spaceflight is associated with increased exposure to various forms of radiation, and even at subclinical exposures, physiological changes take place that affect normal cell, tissue, and organ function (Cucinotta 2008b; Huff et al. 2008; Wu et al. 2008). Direct DNA damage, and the carcinomas that may ensue, are one obvious concern. But even in the absence of DNA damage, an overall increase in oxidative products, including the particularly dangerous reactive oxygen species, seems to occur in the body (Lehnert and Iyer 2002; Baqai et al. 2009). An increase in reactive oxygen species not only is a metabolic strain because of the body's homeostatic drive to maintain reduction-oxidation balance, but also because it can initiate signaling pathways involved in inflammation and repair.

Energy associated with radiation exposure can be absorbed by biological molecules, which can leave them in a highly reactive state. Some of the same unstable reactive groups are normally found in cells (in the electron transport chain for production of energy and in lysosomes for protection from microbial attack). Because they are produced endogenously, cellular mechanisms exist to deactivate them and prevent these reactive species from engaging in unwanted activities and damage. However, ionizing radiation can create reactive species in far greater concentrations than are normally produced, overwhelming the cellular inactivation mechanisms.

The wide spectrum of potential radiation targets (many different proteins, involved in many different cellular processes) offers many therapeutic possibilities, some of which are reviewed in Xiao and Whitnall (2009) and in Coleman et al. (2004). The NIH recently announced a renewed commitment to development of treatments for acute radiation exposure (NIH 2010), and the findings from this program may be useful to NASA even though its aim is to help victims of accidents or war. One current difficulty is determining which molecular targets are most in need of intervention. Others are the diversity of radiation types that the body may encounter during spaceflight and the use of many ground analogs to model spaceflight radiation exposure.

Another confounding factor is that different species, and even strains, of animals seem to differ in their susceptibility to radiation damage or in their ability to repair damage (Pecaut et al.; Lindsay et al. 2007). All of these factors make it difficult to compare studies to each other and to weigh their conclusions appropriately.

Amelioration of Radiation Damage with Pharmaceuticals

Some of the experiments testing possible therapeutics against radiation-induced damage have not been conducted in the most rigorous or useful fashion. Many utilize "single-point pharmacology," in which a test drug is used at one concentration only, leaving open the possibility that the chosen dose was ineffective for that experiment. It could have been too high, increasing the risk of side effects. On the other hand, the chosen test dose could also have been too low to be measurably effective, leading to the incorrect conclusion that the drug had no effect. It is easy to imagine a scenario in which a single-point pharmacology experiment yields no conclusive data, or even worse, leads to an incorrect conclusion.

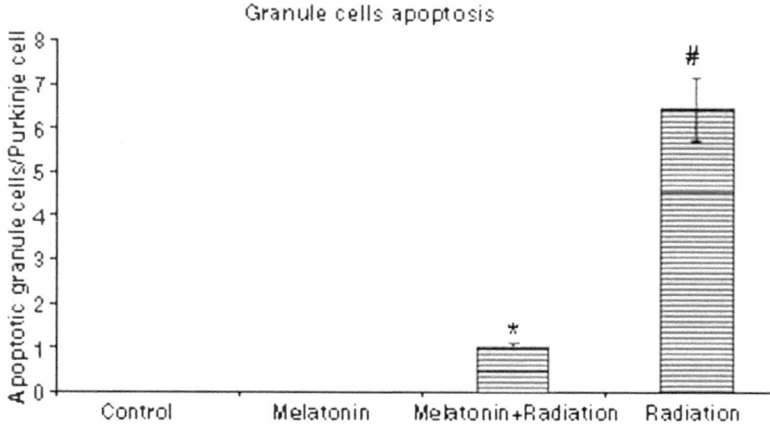

Fig. 9.4 Radiation-induced apoptotic cell death in granule cells with and without melatonin pretreatment. Y-error bars indicate SEM (Manda et al. 2008a). Used with permission, Wiley

For these kinds of reasons, experiments using a single concentration of a drug are not considered useful. Typically, even pilot pharmacology experiments will include at least three concentrations of drug. This maximizes the possibility that the experimental results will be useful. Studies in which a drug is used at only a single dose or concentration must be considered preliminary at best. Given the problems comparing studies across different species and strain, and different radiation types and exposures, in addition to single-point pharmacology issues, the evidence below must be weighed carefully.

Several reducing agents have been tested as pharmacological treatments for radiation-induced damage. A scavenger of oxygen free-radical species, α-lipoic acid (Bilska and Wlodek 2005), has been tested in several scenarios. Mice (male C3H, 8 weeks old) were exposed to an X-ray source (4 or 6 Gy at 0.55 Gy/min), and various measurements were made of reactive oxygen species. A single concentration of α-lipoic acid was found to reduce radiation-induced changes (Manda et al. 2007). In another study by the same laboratory, mice (male C57BL, 8 weeks old) were exposed to high-LET [56]Fe at 1.5 Gy and behavioral as well as cellular and histological analyses were performed on controls and on animals given 200 mg/kg α-lipoic acid (Fig. 9.4). They showed radiation-induced effects on learning behaviors, and on brain cell histology as well as several biochemical measures, all of which were reduced by the α-lipoic acid treatment (Manda et al. 2008b). Another natural plant product, β-carotene, is thought to be a potent antioxidant and has been tested as a radiation protectant. Mice (Swiss albino, 6–8 weeks old) were exposed to 5 Gy γ irradiation, which increased markers of lipid peroxidation and glutathione. These increases were largely prevented by 2 weeks of oral administration of 30 mg/kg β-carotene (Manda and Bhatia 2003).

High-energy proton exposure (1 GeV/n iron ion beam at 40–200 cGy/min) was used to induce oxidative damage in mice (CBA strain, male, 8 weeks old).

Two treatments, each at a single dose, were administered and found to reduce total antioxidant load in the serum: L-selenomethionine (a potent reducing agent) and an antioxidant mixture containing L-selenomethionine, ascorbic acid, N-acetyl cysteine, α-lipoic acid, vitamin E, and coenzyme Q10 (Kennedy et al. 2007). Free radical scavengers have also been shown to reduce damage caused by exposure to 6 Gy of x-irradiation (Murley et al. 2006).

Several naturally produced hormones have been investigated as possible therapeutics against radiation damage. Melatonin, discussed as a potential sleep aid in the CNS section of this work, is by its chemical nature a powerful antioxidant (Reiter et al. 2005; Tan et al. 2007). Ghrelin was first proposed to be a growth hormone regulator and involved in feeding behaviors, but it also seems to be involved in regulation of the immune system and inflammation (Wang et al. 2002; Kojima and Kangawa 2005; Taub 2008). Various steroid hormones can similarly influence activity of the immune system (Whitnall et al. 2005).

Members of the 5-androstene family of steroids have been shown to improve survival of mice after exposure to γ-radiation, likely by stimulation of the immune system (Whitnall et al. 2001; Whitnall et al. 2005), and similar improvements in survival have been shown in 5-androstenediol-treated monkeys exposed to 6 Gy ^{60}Co (Stickney et al. 2007). Results of a study of the effects of a variety of natural and synthetic hormones on γ-irradiated mice suggested that neither androgen nor estrogen receptors are involved in the improved survival (Whitnall et al. 2005). A study on rats (Sprague–Dawley, male, 300 g) exposed to 5 Gy ^{137}Cs and given a septic challenge showed that ghrelin (30 nmol delivered over 3 days) significantly improved survival rate, as well as many biochemical variables (Shah et al. 2009). Melatonin (10 mg/kg) has been shown to ameliorate loss of neurogenesis induced in mice (C57, male, 8 weeks old) by x-irradiation (6 Gy) (Manda et al. 2009). Mice (C57BL, male, 8 weeks old) irradiated with 2 Gy ^{56}Fe showed significantly less apoptosis in cerebellar granule cells after 10 mg/kg melatonin, as shown in Fig. 9.4 (Manda et al. 2008a). Melatonin (0.1 mg/kg) also improved survival of mice (Swiss albino, male, 6–8 weeks old) after γ-irradiation (Bhatia and Manda 2004).

One of melatonin's metabolites has also been shown to increase survival of brain cells after radiation treatment (Manda et al. 2008a). Melatonin has also been shown to improve fluidity in radiation-damaged lipid membranes (Karbownik and Reiter 2000). Genistein, a phytoestrogen derived from soy, at 160 mg/kg improved survival in mice exposed to 6 Gy γ-irradiation, as compared to vehicle control (Zhou and Mi 2005) and at doses of 25–400 mg/kg increases survival of mice exposed to 9.5 Gy γ-irradiation (Landauer et al. 2003). This second study included not only dose–response data but also showed that these genistein doses were well-tolerated by non-irradiated control animals (Landauer et al. 2003).

Many of the proposed treatments for amelioration or prevention of radiation damage are derived from traditional medicine, nutritional supplements, or plant extracts, and as such, fall outside the realm of FDA-regulated pharmaceuticals. They have not been through safety and efficacy testing, and standards do not exist for amounts of active ingredient present per dose. Some intriguing results have been reported in the literature for such therapies, but in most cases, the potency and dose of the active ingredient (and perhaps the identity of the active ingredient) are not known.

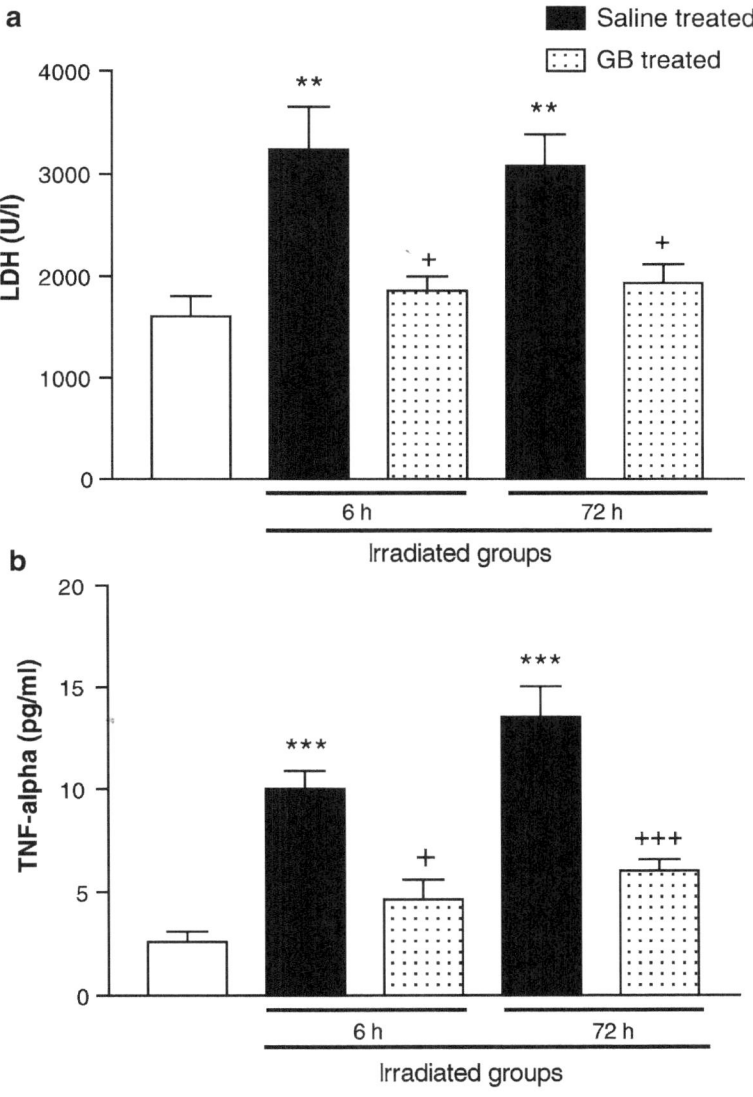

Fig. 9.5 Lactate dehydrogenase (LDH) (*upper*) and TNFα concentrations in control, vehicle, or ginkgo extract (EGb)-treated rats at 6 or 72 h after irradiation with 800 cGy (Sener et al. 2006). Used with permission, Elsevier

Ginkgo biloba is one of these and has been suggested to have antioxidant and free-radical scavenging properties. It was tested as a protectant in rats irradiated at 800 cGy and was found to significantly alter the amounts of several molecules involved in tissue damage, inflammation, and liver function (Fig. 9.5 and Table 9.1) (Sener et al. 2006). It is unfortunate that this report does not provide details of the

Table 9.1 Serum aspartate aminotransferase (AST), alanine aminotransferase (ALT), blood urea nitrogen (BUN), creatinine and lactate dehydrogenase (LDH) concentrations in control, vehicle or ginkgo extract (EGb) treated rats at 6 or 72 h after irradiation with 800 cGy (Sener et al. 2006). Used with permission, Elsevier

		IR groups			
		6 (h)		72 (h)	
	Control	Saline-treated	EGb-treated	Saline-treated	EGb-treated
ALT (U/L)	108 ± 8.9	203 ± 18.3*	113 ± 18.1	235 ± 30.3**	117 ± 16.2+
AST (U/L)	233 ± 29.2	282 ± 24.1	237 ± 26.8	389 ± 46.8*	252 ± 12.5++
BUN (U/L)	34.4 ± 1.7	37.2 ± 2.7	36.2 ± 1.6	60.2 ± 5.3***	36.4 ± 2.6+++
Creatinine (U/L)	0.50 ± 0.02	0.54 ± 0.02	0.58 ± 0.05	0.46 ± 0.03	0.43 ± 0.06

Each group consists of six rats

$*p < 0.05$; $**p < 0.01$; $***p < 0.001$ compared with control group

$+p < 0.01$; $++p < 0.05$; $+++p < 0.001$ compared with saline-treated irradiated group

ginkgo extract that was used. Flaxseed oil has been used in similar experiments, and was shown to significantly improve survival rates as well as biochemical variables, including aspartate aminotransferase and alanine aminotransferase, in mice exposed to 5 Gy (Bhatia et al. 2007). Rosemary extract also showed some improvements in lipid peroxidation and glutathione concentrations of mice after γ-irradiation (Soyal et al. 2007).

Radiation Damage to Stored Pharmaceuticals

Another concern is the possible effect of radiation on both active and inactive ingredients in pharmaceuticals. The inactive ingredients are sometimes considered secondary, but many of these are chosen specifically to allow a particular release profile over time or to lend a certain solubility profile, both extremely important for ultimate bioavailability of the active ingredients. This has been studied in the context of γ-irradiation used for sterilization purposes, typically using ^{60}Co or ^{137}Cs sources at 25 kGy (estimated to be several orders of magnitude more than experienced during flight). Faster dissolution and release have been demonstrated in terrestrial tests with diltiazem in two different excipients (Figs. 9.6 and 9.7) (Maggi et al. 2003). Polyethylene oxide excipients have also been shown to be very sensitive to radiation, and the resulting breakdown results in reduced dissolution and release control (Maggi et al. 2004). If the spaceflight radiation environment causes similar changes in release of active compounds, this would be worth considering during drug selection. Furthermore, radiosterilization has been shown to increase proportion of stereoisomers of the active ingredient found in a cephalosporin antibiotic (Crucq and Tilquin 1996; Barbarin et al. 2001). The ultimate therapeutic effect of this effect is not yet known. An early study showed that the degradation products produced by irradiation were the same as the degradation products seen when the antibiotics were aged in dry heat (Tsuji et al. 1983), but it is not known if this result will hold true for other drug classes.

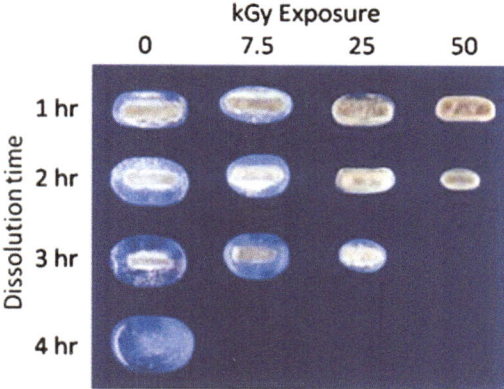

Fig. 9.6 Photograph of tablets during dissolution test (Maggi et al. 2003). Used with permission, Wiley

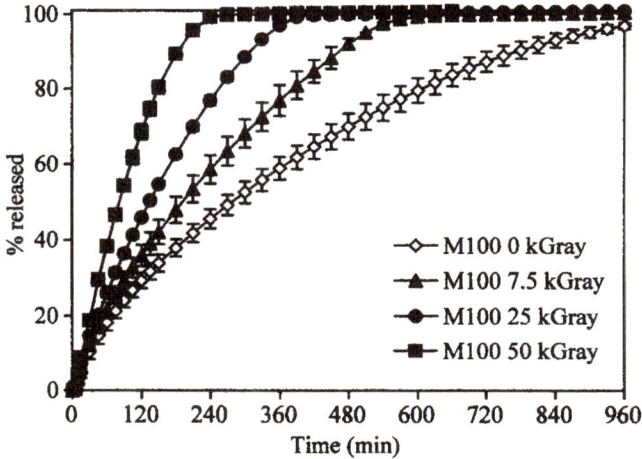

Fig. 9.7 Release of active ingredient over time after indicated radiation exposures (Maggi et al. 2003). Used with permission, Wiley

A recent pilot experiment evaluated the stability of about 30 pharmaceuticals from the ISS medical kit during a flight period of about 2.5 years. Onboard environmental monitoring indicated that temperature and humidity were relatively constant and within normal ranges, leaving radiation as the likely reason for any changes seen in the evaluated medications. On the whole, the drugs tested met U. S. Pharmacopeia (USP) requirements until near the time of their labeled expiration dates. There were six exceptions that decayed earlier than their expiration dates: the antimicrobials Augmentin, levofloxacin, trimethoprim, and sulfamethoxazole, the diuretic furosemide, and the synthetic hormone levothyroxin.

In this study only a single sample of each drug was analyzed, so it is not possible to assign significance to these preliminary results, but they do indicate that at least some drugs in the flight medical kit (as they are currently stowed) merit further attention. Also, only the content of active ingredient was measured from each sample; no screening for appearance of (or toxicity of) breakdown products was performed. In most cases, however, NASA practice is to remove drugs from their manufacturer's packaging for stowage on the ISS; this is done to conserve both mass and volume, but it adds a possible confounding factor. It is possible that the effects seen in this pilot study were caused by poor storage conditions and had little to do with the spaceflight environment or radiation exposure (Du et al. 2011).

Multiple Systems Spaceflight Effects Summary

The evidence regarding immune system responses during spaceflight falls under Category II (controlled), whereas the data from cell cultures is Category I (controlled and randomized). Testing the efficacy of the current anti-infective selections against spaceflight-altered microorganisms should be considered urgent. These are simple studies to conduct that will establish if our current medications are adequate, or if they need to be reconsidered. Ineffective antibiotic treatment of infections could lead to impaired performance or loss of life. Additionally, antibiotic therapy could lead to an altered or compromised GI flora that permits unusual microbial growth.

The evidence cited above for effects on human physiology falls under evidence Category II and III, with some animals studies in Category I (but these are missing dose–response data). The evidence regarding drug stability after radiation exposure falls under Category I, but many more pharmaceuticals remain to be tested under many more radiation exposure scenarios. Additional stability studies should be conducted for the six drugs identified in the pilot study. Ground-based studies to determine the extent of any re-packaging effect should be considered.

Chapter 10
Conclusions: Special Challenges of Long Duration Exploration

With the shuttle making frequent visits to the ISS, re-stocking with fresh supplies, including medications, was routine. With the end of the space shuttle, NASA is now in a transition phase. But in the interest of true exploration, longer missions are under consideration for the future. These may last up to 3 years and re-supply may not be possible. We are taking this opportunity to think ahead to what these very different kinds of missions would require.

Current Drug Testing for Long Duration Spaceflight

Spaceflight missions have unique medical requirements, with demands that increase with mission duration. The FDA drug testing and approval process was not designed to address issues associated with extended storage or exposure to the unique environmental conditions of spaceflight, which means that the additional testing required to ensure safety and efficacy on long-term missions falls to NASA. A comprehensive drug testing plan will be required for mission preparation, a plan that incorporates the research required to answer the questions that are unique to spaceflight.

In planning for future missions, considerations must be addressed that are not examined in the terrestrial drug approval system. Mission duration affects the drugs themselves, as well as the physiological status of the crewmembers, and further investigations are required in both areas to fill in gaps in knowledge. Longer mission duration requires that more attention be paid to shelf life, storage conditions, and the materials used for packaging drugs for extended periods of time.

Certain duties that crewmembers perform require leaving the relative safety of the spacecraft, such as extravehicular activities (EVAs or space walks). The impenetrable suit designed to protect the crewmember from the harsh environment of space puts severe limitations on access to water, food, and also medication administration. Since some EVAs are designed to last for 8 h or longer, there are plans for suit designs that incorporate access to water, food, and certain emergency medications. This may require development of novel formulations and dosage forms.

V.E. Wotring, *Space Pharmacology*, SpringerBriefs in Space Development, DOI 10.1007/978-1-4614-3396-5_10, © Virginia E. Wotring 2012

Unique Medical Requirements for Long Duration Spaceflight

Several of space-related physiological changes increase over time (bone loss, for example), and more information about physiology changes over the course of longer missions is required to ensure adequate care throughout a long mission. But duration is not the only new consideration; in order to explore, missions will have to leave the relative protection of Earth's magnetic field, and will thus be exposed to increased amounts of radiation. The exposure to these higher radiation levels would continue over the entire duration of the mission, so that total exposure could be increased quite significantly over what crewmembers currently experience in low-Earth orbit. This applies to both crewmembers and to stored medications. Plans for long duration spaceflight should include a mechanism to consider the addition of new drugs to the medical kit, to prepare for both chronic and acute conditions that may occur in longer duration flights, as well as to prepare for the additional medical events that can be anticipated on any long journey.

Packaging and Shelf Life

Current pharmaceutical industry practice is to supply products in packaging that adequately protects for a period of at least 1 year (if possible) from the light, heat, humidity, and oxygen encountered in ordinary indoor conditions. Terrestrially, it is considered practical to match drug supply with demand, so stability during long-term storage is not a significant issue.

Whatever long-term stability testing is conducted by pharmaceutical companies is the property of the company and is not published. An extended shelf-life study is being conducted by the FDA with the Department of Defense, but they have selected drugs appropriate for military or public emergency scenarios; they have not included many of the drugs that are expected to be required in spaceflight. Collaboration with this study would be helpful, but would not provide information on all of the drugs of interest and would not include any information on storage in the re-packaging system that NASA typically uses to reduce mass and volume and waste. In order to know the actual shelf life of the components of the flight kit as flown, NASA will have to begin these studies soon. Concentration of active ingredients and formula bioavailability should be measured over time. Another critical component of these stability studies is examination of any breakdown products for toxicity. Those medications that exhibit significant breakdown over time, or whose breakdown products are toxic, would not be recommended for inclusion in flight medical kits. The results of such a study could mean that storage conditions during missions need to be altered.

Currently, NASA re-packages many pharmaceuticals into packaging that is operationally convenient (low mass, low volume, no packaging waste), but it is not yet known how this re-packaging affects the shelf life of the stored drugs. This testing must be completed before long term missions begin so that storage conditions can be improved if needed. Since additional shielding against radiation or other environmental factors may also be desirable on longer missions, testing of packaging systems that incorporate some level of shielding would be advisable. Systematic studies to test the stability, bioavailability, and production of toxic metabolites of drugs considered for inclusion in flight kits would then permit informed recommendations regarding packaging systems.

It is clear that much remains to done before the effects of pharmaceuticals during spaceflight are understood well enough to make confident predictions of what will occur when medications are used during long duration spaceflights. It has been known for decades that many physiological systems are altered (some dramatically) by the spaceflight environment, although in most cases the details and mechanisms still have yet to be determined. Since these physiological systems are the basis for therapeutic action on the body, any changes they undergo may result in altered pharmaceutical performance. This forms the basis of the molecular mechanism of action of every administered medication, and at present, our knowledge is based largely on case reports from very few individuals and anecdotal evidence from a few more. Additional details are required for physiological changes that depend on mission length. Bone loss, muscle atrophy, and radiation exposure effects all seem to seem to increase with mission duration, but it is not yet known if they continue to change at constant rates during extended missions, or if the effects diminish over time, or if homeostatic mechanisms can return the system to normal. For other systems, such as the immune system, it simply is not yet known if physiological changes stabilize to a new "space normal" or if they continue to accrue over time.

All of these unknowns in the physiology affect our ability to predict how effective or safe administered medications may prove to be. Furthermore, extensive drug storage testing needs to be completed in order to prepare for long range exploratory missions. Changes in packaging materials and methods may be required to better preserve medications for longer periods of time. In order to best protect and treat crewmembers on longer, more exploratory missions, there are many pharmacological questions that must be answered.

Abbreviations

ACTH	Adrenocorticotropic hormone
ADH	Antidiuretic hormone, also known as vasopressin or arginine vasopressin
ADHD	Attention deficit hyperactivity disorder
ADME	Absorption, distribution, metabolism, excretion; the four components of pharmacokinetics
ANP	Atrial naturetic peptide
ATP	Adenosine triphosphate
AVP	Arginine vasopressin
AUC	Area under the curve, a measure of total drug in the system integrated over time
BHP	Behavioral Health Program
BMD	Bone mineral density
cAMP	Cyclic adenosine monophosphate
CNS	Central nervous system
CTZ	Chemoreceptive trigger zone
CV	Cardiovascular
CYP	Name for individual enzyme in the P450 family, the number following designates the specific enzyme and isoform
Cyt P450	Cytochrome P450 enzymes
d	Day(s)
D1	Dopamine receptor type 1
D2	Dopamine receptor type 2
DA	Dopamine
DIGE	Difference gel electrophoresis (a quantitative 2D method)
DNA	Deoxyribonucleic acid
DUI	Driving under the influence
EEG	Electroencephologram
EVA	Extravehicular activity (spacewalk)

V.E. Wotring, *Space Pharmacology*, SpringerBriefs in Space Development, DOI 10.1007/978-1-4614-3396-5, © Virginia E. Wotring 2012

FDA   Food and drug administration

G   Gravitational force
GABA γ-aminobutyric acid
GABA$_A$R γ-aminobutyric acid receptor type A
GI   Gastrointestinal
GR   205171 a neurokinin antagonist

^{3}H   Tritium, a radioactive form of hydrogen used as a tracer
H1   Histamine receptor type 1
HDT   Head-down tilt
5HT$_3$ Serotonin type 3 receptor

IM   Intramuscular
IN   Intranasal
ISS   International Space Station
IV   Intravenous

Kg   Kilogram

L   Liter
L-758,298   A neurokinin antagonist
LGD 2226   A selective androgen receptor modulator

mg   Milligram
mRNA   Messenger ribonucleic acid

n   The number of biological samples in an experiment
NE   Norepinephrine
NK   Neurokinin

OH   Orthostatic hypotension
OI   Orthostatic intolerance

PD   Pharmcodynamics, the study of drug action on the body
pH   Unit of acidity (inverse log of hydronium ion concentration)
PK   Pharmacokinetics, the study of how the body acts upon a drug after it is administered
POI   Postflight orthostatic intolerance
PVT   Psychomotor vigilance test
PO   *Per os* (by mouth, or oral drug administration)
PR   *Per rectum* (rectal drug administration)

QT   Time between the start of the Q wave and the end of the T wave in the heart's electrical cycle

RANK   Receptor activator for nuclear factor κ B
RANKL   Receptor activator for nuclear factor κ B ligand
REM   Rapid eye movement, a sleep phase
RNA   Ribonucleic acid

SARM	Selective androgen receptor modulator
SMS	Space motion sickness
SPARC	Gene that codes for osteonectin
STS	Space transportation system (space shuttle)
TD	Transdermal (drug administration using a patch applied to the skin)
UPA	Urine process assembly
VO$_2$ max	Maximum oxygen consumption

References

S. Abraham, C.Y. Lin et al., Studies of specific hepatic enzymes involved in the conversion of carbohydrates to lipids in rats exposed to prolonged spaceflight aboard Cosmos 1129. Physiologist **23**(Suppl 6), S55–S58 (1980)

W.R. Adey, Studies on weightlessness in a primate in the Biosatellite 3 experiment. Life Sci. Space Res. **10**, 67–85 (1972)

R. Agarwal, Regulation of circadian blood pressure: from mice to astronauts. Curr. Opin. Nephrol. Hypertens. **19**(1), 51–58 (2010)

A.F.H.R.A. Quality. Medical Expenditure Panel Survey, 2010, from http://www.ahrq.gov/data/mepsix.htm. (2009)

Z. Allebban, L.A. Gibson et al., Effects of spaceflight on rat erythroid parameters. J. Appl. Physiol. **81**(1), 117–122 (1996)

B. Ameer, M. Divoll et al., Absolute and relative bioavailability of oral acetaminophen preparations. J. Pharm. Sci. **72**(8), 955–958 (1983)

P.L. Andrews, F. Okada et al., The emetic and anti-emetic effects of the capsaicin analogue resiniferatoxin in Suncus murinus, the house musk shrew. Br. J. Pharmacol. **130**(6), 1247–1254 (2000)

R.H. Anken, R. Hilbig, A drop-tower experiment to determine the threshold of gravity for inducing motion sickness in fish. Adv. Space Res. **34**(7), 1592–1597 (2004)

J. Arendt, M. Aldhous et al., Alleviation of jet lag by melatonin: preliminary results of controlled double blind trial. Br. Med. J. (Clin. Res. Ed.) **292**(6529), 1170 (1986)

J. Arendt, V. Marks, Jet lag and melatonin. Lancet **2**(8508), 698–699 (1986)

J.P. Bagian, D.F. Ward, A retrospective study of promethazine and its failure to produce the expected incidence of sedation during space flight. J. Clin. Pharmacol. **34**(6), 649–651 (1994)

Y.D. Bai, F.S. Yang et al., Inhibition of RANK/RANKL signal transduction pathway: a promising approach for osteoporosis treatment. Med. Hypotheses **71**(2), 256–258 (2008)

E. Balaban, C. Centini et al., Tonic gravity changes alter gene expression in the efferent vestibular nucleus. Neuroreport **13**(1), 187–190 (2002)

S. Banks, D.F. Dinges, Behavioral and physiological consequences of sleep restriction. J. Clin. Sleep Med. **3**(5), 519–528 (2007)

G.R. Banta, W.C. Ridley et al., Aerobic fitness and susceptibility to motion sickness. Aviat. Space Environ. Med. **58**(2), 105–108 (1987)

F.P. Baqai, D.S. Gridley et al., Effects of spaceflight on innate immune function and antioxidant gene expression. J. Appl. Physiol. **106**(6), 1935–1942 (2009)

N. Barbarin, B. Tilquin et al., Radiosterilization of cefotaxime: investigation of potential degradation compounds by liquid chromatography-electrospray mass spectrometry. J. Chromatogr. A **929**(1–2), 51–61 (2001)

M. Basner, J. Rubinstein et al., Effects of night work, sleep loss and time on task on simulated threat detection performance. Sleep **31**(9), 1251–1259 (2008)

M.A. Bayorh, R.R. Socci et al., L-NAME, a nitric oxide synthase inhibitor, as a potential counter-measure to post-suspension hypotension in rats. Clin. Exp. Hypertens. **23**(8), 611–622 (2001)

M. Beaumont, D. Batejat et al., Zaleplon and zolpidem objectively alleviate sleep disturbances in mountaineers at a 3,613 meter altitude. Sleep **30**(11), 1527–1533 (2007)

P.M. Becker, M. Sattar, Treatment of sleep dysfunction and psychiatric disorders. Curr. Treat. Options Neurol. **11**(5), 349–357 (2009)

Z. Ben-Zvi, E. Gussarsky et al., The bioavailability of febantel in dehydrated camels. J. Vet. Pharmacol. Ther. **19**(4), 288–294 (1996)

I. Berlin, D. Warot et al., Comparison of the effects of zolpidem and triazolam on memory functions, psychomotor performances, and postural sway in healthy subjects. J. Clin. Psychopharmacol. **13**(2), 100–106 (1993)

A.T. Bernardini, M. Taub, Effects of reduced pressure and drug administration on the glucose tolerance test in the dog. Aerosp. Med. **40**(4), 409–412 (1969)

R. Berne, M. Levy, *Physiology* (Mosby, St. Louis, 1988)

S. Bhasin, O.M. Calof et al., Drug insight: testosterone and selective androgen receptor modulators as anabolic therapies for chronic illness and aging. Nat. Clin. Pract. Endocrinol. Metab. **2**(3), 146–159 (2006)

S. Bhasin, R. Jasuja, Selective androgen receptor modulators as function promoting therapies. Curr. Opin. Clin. Nutr. Metab. Care **12**(3), 232–240 (2009)

S. Bhasin, T.W. Storer et al., The effects of supraphysiologic doses of testosterone on muscle size and strength in normal men. N. Engl. J. Med. **335**(1), 1–7 (1996)

S. Bhasin, T.W. Storer et al., Testosterone replacement increases fat-free mass and muscle size in hypogonadal men. J. Clin. Endocrinol. Metab. **82**(2), 407–413 (1997)

S. Bhasin, L. Woodhouse et al., Testosterone dose–response relationships in healthy young men. Am. J. Physiol. Endocrinol. Metab. **281**(6), E1172–E1181 (2001)

S. Bhasin, L. Woodhouse et al., Older men are as responsive as young men to the anabolic effects of graded doses of testosterone on the skeletal muscle. J. Clin. Endocrinol. Metab. **90**(2), 678–688 (2005)

A.L. Bhatia, K. Manda, Study on pretreatment of melatonin against radiation-induced oxidative stress in mice. Environ. Toxicol. Pharmacol. **18**, 13–20 (2004)

A.L. Bhatia, A. Sharma et al., Prophylactic effect of flaxseed oil against radiation-induced hepato-toxicity in mice. Phytother. Res. **21**(9), 852–859 (2007)

A. Bilska, L. Wlodek, Lipoic acid – the drug of the future? Pharmacol. Rep. **57**(5), 570–577 (2005)

D.M. Black, D.E. Thompson et al., Fracture risk reduction with alendronate in women with osteo-porosis: the Fracture Intervention Trial. FIT Research Group. J. Clin. Endocrinol. Metab. **85**(11), 4118–4124 (2000)

W. Bles, B. de Graaf et al., A sustained hyper-g load as a tool to simulate space sickness. J. Gravit. Physiol. **4**(2), P1–P4 (1997)

M.H. Bonnet, T.J. Balkin et al., The use of stimulants to modify performance during sleep loss: a review by the sleep deprivation and Stimulant Task Force of the American Academy of Sleep Medicine. Sleep **28**(9), 1163–1187 (2005)

H.L. Borison, A 1983 neuropharmacologic perspective of space sickness. Brain Behav. Evol. **23**(1–2), 7–13 (1983)

J.E. Bos, W. Bles et al., Eye movements to yaw, pitch, and roll about vertical and horizontal axes: adaptation and motion sickness. Aviat. Space Environ. Med. **73**(5), 436–444 (2002)

D.D.A. Bourne (2001), http://www.boomer.org/c/p1/Ch18/Ch1802.html

J.L. Boyd, B. Du et al., Relative bioavailability of scopolamine dosage forms and interaction with dextroamphetamine. J. Gravit. Physiol. **14**(1), P107–P108 (2007)

A. Boyum, P. Wiik et al., The effect of strenuous exercise, calorie deficiency and sleep deprivation on white blood cells, plasma immunoglobulins and cytokines. Scand. J. Immunol. **43**(2), 228–235 (1996)

K.T. Brixen, P.M. Christensen et al., Teriparatide (biosynthetic human parathyroid hormone 1–34): a new paradigm in the treatment of osteoporosis. Basic Clin. Pharmacol. Toxicol. **94**(6), 260–270 (2004)

J. Broskey, M.K. Sharp, Evaluation of mechanisms of postflight orthostatic intolerance with a simple cardiovascular system model. Ann. Biomed. Eng. **35**(10), 1800–1811 (2007)

T.E. Brown, D.L. Eckberg, Promethazine affects autonomic cardiovascular mechanisms minimally. J. Pharmacol. Exp. Ther. **282**(2), 839–844 (1997)

L.J. Brunner, S. Bai et al., Effect of simulated weightlessness on phase II drug metabolism in the rat. Aviat. Space Environ. Med. **71**(9), 899–903 (2000)

L.J. Brunner, J.T. DiPiro et al., Antipyrine pharmacokinetics in the tail-suspended rat model. J. Pharmacol. Exp. Ther. **274**(1), 345–352 (1995)

J.C. Buckey, *Space Physiology* (Oxford, New York, 2006)

J.C. Buckey, D. Alvarenga et al., Chlorpheniramine for motion sickness. J. Vestib. Res. **14**(1), 53–61 (2004)

J.C. Buckey Jr., L.D. Lane et al., Orthostatic intolerance after spaceflight. J. Appl. Physiol. **81**(1), 7–18 (1996)

M.W. Bungo, J.B. Charles et al., Cardiovascular deconditioning during space flight and the use of saline as a countermeasure to orthostatic intolerance. Aviat. Space Environ. Med. **56**(10), 985–990 (1985)

S.A. Bustin, Why the need for qPCR publication guidelines? – The case for MIQE. Methods **50**(4), 217–226 (2010)

S.A. Bustin, V. Benes et al., The MIQE guidelines: minimum information for publication of quantitative real-time PCR experiments. Clin. Chem. **55**(4), 611–622 (2009)

J.A. Caldwell, N.K. Smythe et al., Efficacy of Dexedrine for maintaining aviator performance during 64 hours of sustained wakefulness: a simulator study. Aviat. Space Environ. Med. **71**(1), 7–18 (2000)

C. Carcenac, S. Herbute et al., Hindlimb-suspension and spaceflight both alter cGMP levels in rat choroid plexus. J. Gravit. Physiol. **6**(2), 17–24 (1999)

R. Casaburi, S. Bhasin et al., Effects of testosterone and resistance training in men with chronic obstructive pulmonary disease. Am. J. Respir. Crit. Care Med. **170**(8), 870–878 (2004)

P.R. Cavanagh, A.A. Licata et al., Exercise and pharmacological countermeasures for bone loss during long-duration space flight. Gravit. Space Biol. Bull. **18**(2), 39–58 (2005)

J.B. Charles, M.W. Bungo, Cardiovascular physiology in space flight. Exp. Gerontol. **26**(2–3), 163–168 (1991)

J. Chen, J. Kim et al., Discovery and therapeutic promise of selective androgen receptor modulators. Mol. Interv. **5**(3), 173–188 (2005)

M.M. Chi, R. Choksi et al., Effects of microgravity and tail suspension on enzymes of individual soleus and tibialis anterior fibers. J. Appl. Physiol. **73**(2 Suppl), 66S–73S (1992)

K. Cho, A. Ennaceur et al., Chronic jet lag produces cognitive deficits. J. Neurosci. **20**(6), RC66 (2000)

M.M. Cho, N.P. Ziats et al., Effects of estrogen on tight junctional resistance in cultured human umbilical vein endothelial cells. J. Soc. Gynecol. Investig. **5**(5), 260–270 (1998)

P. Chowdhury, M.E. Soulsby et al., Distribution of 3H-nicotine in rat tissues under the influence of simulated microgravity. Biomed. Environ. Sci. **12**(2), 103–109 (1999)

P. Chowdhury, M. Soulsby, et al., L-carnitine influence on oxidative stress induced by hind limb unloading in adult rats. Aviat. Space Environ. Med. **78**(6), 554–556 (2007)

N.M. Cintron, H.W. Lane et al., Metabolic consequences of fluid shifts induced by microgravity. Physiologist **33**(1 Suppl), S16–S19 (1990)

N.M. Cintron, L. Putcha, et al., In-flight salivary pharmacokinetics of scopalamine and dextramphetamine. NASA Technical Memorandum 58280, Houston, NASA, 1987b, pp. 25–29

N.M. Cintron, L. Putcha, et al., In-flight pharmacokinetics of acetaminophen in saliva. NASA Technical Memorandum 58280, Houston, NASA, 1987a, pp. 19–23

G. Clement, *Fundamentals of Space Medicine* (Microcosm and Kluwer, El Segundo, CA, 2003)

J.A. Clements, R.C. Heading et al., Kinetics of acetaminophen absorption and gastric emptying in man. Clin. Pharmacol. Ther. **24**(4), 420–431 (1978)

S.P. Clissold, R.C. Heel, Transdermal hyoscine (Scopolamine). A preliminary review of its pharmacodynamic properties and therapeutic efficacy. Drugs **29**(3), 189–207 (1985)

B. Cohen, M. Dai et al., Baclofen, motion sickness susceptibility and the neural basis for velocity storage. Prog. Brain Res. **171**, 543–553 (2008)

C.N. Coleman, H.B. Stone et al., Medicine. Modulation of radiation injury. Science **304**(5671), 693–694 (2004)

P.J. Coleman, J.J. Renger, Orexin receptor antagonists: a review of promising compounds patented since 2006. Expert Opin. Ther. Pat. **20**(3), 307–324 (2010)

G.A. Conder, H.S. Sedlacek et al., Efficacy and safety of maropitant, a selective neurokinin 1 receptor antagonist, in two randomized clinical trials for prevention of vomiting due to motion sickness in dogs. J. Vet. Pharmacol. Ther. **31**(6), 528–532 (2008)

J. Connolly, J. Boulter et al., Alpha 4–2 beta 2 and other nicotinic acetylcholine receptor subtypes as targets of psychoactive and addictive drugs. Br. J. Pharmacol. **105**(3), 657–666 (1992)

V.A. Convertino, D.A. Ludwig et al., Effects of exposure to simulated microgravity on neuronal catecholamine release and blood pressure responses to norepinephrine and angiotensin. Clin. Auton. Res. **8**(2), 101–110 (1998)

P. Cowings, C. Stout, et al., The effects of promethazine on human performance, mood states and motion sickness tolerance. NASA Technical Memorandum 110420, (1996)

P.S. Cowings, W.B. Toscano, Autogenic-feedback training exercise is superior to promethazine for control of motion sickness symptoms. J. Clin. Pharmacol. **40**(10), 1154–1165 (2000)

P.S. Cowings, W.B. Toscano et al., Promethazine as a motion sickness treatment: impact on human performance and mood states. Aviat. Space Environ. Med. **71**(10), 1013–1022 (2000)

J.A. Cramer, R.H. Mattson et al., How often is medication taken as prescribed? A novel assessment technique. JAMA **261**(22), 3273–3277 (1989)

G.H. Crampton, J.B. Lucot, A stimulator for laboratory studies of motion sickness in cats. Aviat. Space Environ. Med. **56**(5), 462–465 (1985)

A. Crema, G.M. Frigo et al., A pharmacological analysis of the peristaltic reflex in the isolated colon of the guinea-pig or cat. Br. J. Pharmacol. **39**(2), 334–345 (1970)

B. Crucian, *Risk of Crew Adverce Health Event Due to Altered Immune Response* (NASA, Houston TX, HHC, 2009)

A.S. Crucq, B. Tilquin, Method to identify products induced by radiosterilization. A study of cefotaxime sodium. J. Pharm. Belg. **51**(6), 285–288 (1996)

F.A. Cucinotta, M. Durante, *Risk of Radiation Carcinogenesis* (NASA, Houston TX, HHC, 2008a)

F.A. Cucinotta, H. Wang, et al., *Risk of Acute or Late Central Nervous System Effects from Radiation Exposure* (NASA, Houston TX, HHC, 2008b)

S.R. Cummings, D.M. Black et al., Effect of alendronate on risk of fracture in women with low bone density but without vertebral fractures: results from the Fracture Intervention Trial. JAMA **280**(24), 2077–2082 (1998)

T.R. Czarnik, J. Vernikos, Physiological changes in spaceflight that may affect drug action. J. Gravit. Physiol. **6**(1), P161–P164 (1999)

C.A. Czeisler, A.J. Chiasera et al., Research on sleep, circadian rhythms and aging: applications to manned spaceflight. Exp. Gerontol. **26**(2–3), 217–232 (1991)

C.A. Czeisler, J.J. Gooley, Sleep and circadian rhythms in humans. Cold Spring Harb. Symp. Quant. Biol. **72**, 579–597 (2007)

D.S. D'Aunno, A.H. Dougherty et al., Effect of short- and long-duration spaceflight on QTc intervals in healthy astronauts. Am. J. Cardiol. **91**(4), 494–497 (2003)

V.C. da Silva, P.L. Bittencourt et al., Delayed-onset hepatic encephalopathy induced by zolpidem: a case report. Clinics (Sao Paulo) **63**(4), 565–566 (2008)

J.T. Dalton, A. Mukherjee et al., Discovery of nonsteroidal androgens. Biochem. Biophys. Res. Commun. **244**(1), 1–4 (1998)

P. Danjou, I. Paty et al., A comparison of the residual effects of zaleplon and zolpidem following administration 5 to 2 h before awakening. Br. J. Clin. Pharmacol. **48**(3), 367–374 (1999)

J.R. Davis, R.T. Jennings et al., Comparison of treatment strategies for space motion sickness. Acta Astronaut. **29**(8), 587–591 (1993a)

J.R. Davis, R.T. Jennings et al., Treatment efficacy of intramuscular promethazine for space motion sickness. Aviat. Space Environ. Med. **64**(3 Pt 1), 230–233 (1993b)

J.R. Davis, J.M. Vanderploeg et al., Space motion sickness during 24 flights of the space shuttle. Aviat. Space Environ. Med. **59**(12), 1185–1189 (1988)

L.R. Davrath, R.W. Gotshall et al., Moderate sodium restriction does not alter lower body negative pressure tolerance. Aviat. Space Environ. Med. **70**(6), 577–582 (1999)

L. Debuys, A. Henrique, Effect of body posture on the position and emptying time of the stomach. Am. J. Dis. Child. **15**, 190 (1918)

A. Descamps, C. Rousset et al., Influence of the novel antidepressant and melatonin agonist/sero-tonin2C receptor antagonist, agomelatine, on the rat sleep-wake cycle architecture. Psychopharmacology (Berl.) **205**(1), 93–106 (2009)

S.G. Diamond, C.H. Markham, Otolith function in hypo- and hypergravity: relation to space motion sickness. Acta Otolaryngol. Suppl. **481**, 19–22 (1991)

A. Diedrich, S.Y. Paranjape et al., Plasma and blood volume in space. Am. J. Med. Sci. **334**(1), 80–85 (2007)

D.J. Dijk, D.F. Neri et al., Sleep, performance, circadian rhythms, and light–dark cycles during two space shuttle flights. Am. J. Physiol. Regul. Integr. Comp. Physiol. **281**(5), R1647–R1664 (2001)

D.F. Dinges, Sleep in space flight: breath easy–sleep less? Am. J. Respir. Crit. Care Med. **164**(3), 337–338 (2001)

D.F. Dinges, S. Arora et al., Pharmacodynamic effects on alertness of single doses of armodafinil in healthy subjects during a nocturnal period of acute sleep loss. Curr. Med. Res. Opin. **22**(1), 159–167 (2006)

D.F. Dinges, S.D. Douglas et al., Leukocytosis and natural killer cell function parallel neurobehav-ioral fatigue induced by 64 hours of sleep deprivation. J. Clin. Invest. **93**(5), 1930–1939 (1994)

J.M. Discipline, *Risk of Impaired Performance Due to Reduced Muscle Mass, Strength and Endurance* (NASA, Houston TX, HHC, 2008)

J.B. Dressman, P. Bass et al., Gastrointestinal parameters that influence oral medications. J. Pharm. Sci. **82**(9), 857–872 (1993)

C. Drummer, R. Gerzer et al., Body fluid regulation in micro-gravity differs from that on Earth: an overview. Pflugers Arch. **441**(2–3 Suppl), R66–R72 (2000a)

C. Drummer, M. Heer et al., Reduced natriuresis during weightlessness. Clin. Investig. **71**, 678–686 (1993)

C. Drummer, C. Hesse et al., Water and sodium balances and their relation to body mass changes in microgravity. Eur. J. Clin. Invest. **30**(12), 1066–1075 (2000b)

C. Drummer, P. Norsk et al., Water and sodium balance in space. Am. J. Kidney Dis. **38**(3), 684–690 (2001)

S.P. Drummond, A. Bischoff-Grethe et al., The neural basis of the psychomotor vigilance task. Sleep **28**(9), 1059–1068 (2005)

B. Du, V. Daniels et al., Evaluation of physical and chemical changes in pharmaceuticals flown on space missions. AAPS J. **13**(2), 299–308 (2011)

J.P. Edwards, S.J. West et al., New nonsteroidal androgen receptor modulators based on 4-(trifluoromethyl)-2(1H)-pyrrolidino[3,2-g] quinolinone. Bioorg. Med. Chem. Lett. **8**(7), 745–750 (1998)

A.R. Elliott, S.A. Shea et al., Microgravity reduces sleep-disordered breathing in humans. Am. J. Respir. Crit. Care Med. **164**(3), 478–485 (2001)

H.A. Elsheikh, A.M. Osman Intisar et al., Effect of dehydration on the pharmacokinetics of oxytet-racycline hydrochloride administered intravenously in goats (Capra hircus). Gen. Pharmacol. **31**(3), 455–458 (1998)

M. Erman, D. Seiden et al., An efficacy, safety, and dose–response study of Ramelteon in patients with chronic primary insomnia. Sleep Med. **7**(1), 17–24 (2006)

E. Faugloire, C.T. Bonnet et al., Motion sickness, body movement, and claustrophobia during pas-sive restraint. Exp. Brain Res. **177**(4), 520–532 (2007)

FDA (2008), FDA-approved manufacturer drug insert: drug specific medication guides. *NDA 19908 FDA SLR 027 Approved labeling 4.23.08,* http://www.fda.gov/downloads/Drugs/DrugSafety/ucm085906.pdf

FDA, Midodrine hydrochloride: FDA proposes withdrawal of low blood pressure drug (2010), http://www.fda.gov/Safety/MedWatch/SafetyInformation/SafetyAlertsforHumanMedicalProducts/ucm222640.htm

C. Fernandez, J.R. Lindsay, The vestibular coriolis reaction. Arch. Otolaryngol. **80**, 469–472 (1964)

L. Ferraro, T. Antonelli et al., Modafinil: an antinarcoleptic drug with a different neurochemical profile to d-amphetamine and dopamine uptake blockers. Biol. Psychiatry **42**(12), 1181–1183 (1997)

D. Fleisher, C. Li et al., Drug, meal and formulation interactions influencing drug absorption after oral administration. Clinical implications. Clin. Pharmacokinet. **36**(3), 233–254 (1999)

S. Folkard, J. Arendt et al., Melatonin stabilises sleep onset time in a blind man without entrainment of cortisol or temperature rhythms. Neurosci. Lett. **113**(2), 193–198 (1990)

M.A. Frey, J. Riddle et al., Blood and urine responses to ingesting fluids of various salt and glucose concentrations. J. Clin. Pharmacol. **31**(10), 880–887 (1991)

J.M. Fritsch-Yelle, U.A. Leuenberger et al., An episode of ventricular tachycardia during long-duration spaceflight. Am. J. Cardiol. **81**(11), 1391–1392 (1998)

J.D. Frost Jr., W.H. Shumate et al., The Skylab sleep monitoring experiment: methodology and initial results. Acta Astronaut. **2**(3–4), 319–336 (1975)

R.L. Fucci, J. Gardner et al., Toward optimizing lighting as a countermeasure to sleep and circadian disruption in space flight. Acta Astronaut. **56**(9–12), 1017–1024 (2005)

V. Fusco, A. Baraldi et al., Jaw osteonecrosis associated with intravenous bisphosphonate: is incidence reduced after adoption of dental preventive measures? J. Oral. Maxillofac. Surg. **67**(8), 1775 (2009)

P. Gandia, M.P. Bareille et al., Influence of simulated weightlessness on the oral pharmacokinetics of acetaminophen as a gastric emptying probe in man: a plasma and a saliva study. J. Clin. Pharmacol. **43**(11), 1235–1243 (2003)

P. Gandia, S. Saivin et al., The influence of weightlessness on pharmacokinetics. Fundam. Clin. Pharmacol. **19**(6), 625–636 (2005)

C.J. Gardner, D.R. Armour et al., GR205171: a novel antagonist with high affinity for the tachykinin NK1 receptor, and potent broad-spectrum anti-emetic activity. Regul. Pept. **65**(1), 45–53 (1996)

O.H. Gauer, J.P. Henry, Circulatory basis of fluid volume control. Physiol. Rev. **43**, 423–481 (1963)

A. Germain, D.J. Buysse et al., Sleep-specific mechanisms underlying posttraumatic stress disorder: integrative review and neurobiological hypotheses. Sleep Med. Rev. **12**(3), 185–195 (2008)

A. Gilman, T.W. Rall et al. (eds.), *The Pharmacological Basis of Therapeutics* (Pergamon Press, New York, 1990)

J. Gisolf, R.V. Immink et al., Orthostatic blood pressure control before and after spaceflight, determined by time-domain baroreflex method. J. Appl. Physiol. **98**(5), 1682–1690 (2005)

R. Gopalakrishnan, K.O. Genc et al., Muscle volume, strength, endurance, and exercise loads during 6-month missions in space. Aviat. Space Environ. Med. **81**(2), 91–102 (2010)

S. Grady, D. Aeschbach et al., Effect of modafinil on impairments in neurobehavioral performance and learning associated with extended wakefulness and circadian misalignment. Neuropsychopharmacology **35**(9), 1910–1920 (2010)

A. Graybiel, The prevention of motion sickness in orbital flight. Life Sci. Space Res. **14**, 109–118 (1976)

A. Graybiel, J. Knepton, Sopite syndrome: a sometimes sole manifestation of motion sickness. Aviat. Space Environ. Med. **47**(8), 873–882 (1976)

A. Graybiel, C.D. Wood et al., Human assay of antimotion sickness drugs. Aviat. Space Environ. Med. **46**(9), 1107–1118 (1975)

S.A. Greenacre, H. Ischiropoulos, Tyrosine nitration: localisation, quantification, consequences for protein function and signal transduction. Free Radic. Res. **34**(6), 541–581 (2001)

D.J. Greenblatt, J.S. Harmatz et al., Comparative kinetics and dynamics of zaleplon, zolpidem, and placebo. Clin. Pharmacol. Ther. **64**(5), 553–561 (1998a)

D.J. Greenblatt, L.L. von Moltke et al., Kinetic and dynamic interaction study of zolpidem with ketoconazole, itraconazole, and fluconazole. Clin. Pharmacol. Ther. **64**(6), 661–671 (1998b)

D. Grundy, K. Reid et al., Trans-thoracic fluid shifts and endocrine responses to 6 degrees head-down tilt. Aviat. Space Environ. Med. **62**(10), 923–929 (1991)

A. Gundel, V. Nalishiti et al., Sleep and circadian rhythm during a short space mission. Clin. Investig. **71**(9), 718–724 (1993)

A. Gundel, V.V. Polyakov et al., The alteration of human sleep and circadian rhythms during spaceflight. J. Sleep Res. **6**(1), 1–8 (1997)

A. Guyton, J. Hall, *Textbook of Medical Physiology* (Elsevier Saunders, Philadelphia, 2006)

R.R. Hajjar, F.E. Kaiser et al., Outcomes of long-term testosterone replacement in older hypogonadal males: a retrospective analysis. J. Clin. Endocrinol. Metab. **82**(11), 3793–3796 (1997)

R. Hardeland, B. Poeggeler et al., Melatonergic drugs in clinical practice. Arzneimittelforschung **58**(1), 1–10 (2008)

A.R. Hargens, S. Richardson, Cardiovascular adaptations, fluid shifts, and countermeasures related to space flight. Respir. Physiol. Neurobiol. **169**(Suppl 1), S30–S33 (2009)

A.R. Hargens, D.E. Watenpaugh, Cardiovascular adaptation to spaceflight. Med. Sci. Sports Exerc. **28**(8), 977–982 (1996)

J.L. Hargrove, D.P. Jones, Hepatic enzyme adaptation in rats after space flight. Physiologist **28**(6 Suppl), S230 (1985)

D.L. Harm, D.E. Parker et al., Relationship between selected orientation rest frame, circular vection and space motion sickness. Brain Res. Bull. **47**(5), 497–501 (1998)

R.P. Heaney, A.J. Yates et al., Bisphosphonate effects and the bone remodeling transient. J. Bone Miner. Res. **12**(8), 1143–1151 (1997)

M. Heer, W.H. Paloski, Space motion sickness: incidence, etiology, and countermeasures. Auton. Neurosci. **129**(1–2), 77–79 (2006)

R.M. Heggie, I.R. Entwistle, Seasickness. Br. Med. J. **4**(5629), 514 (1968)

L. Henriksen, P. Sejrsen, Local reflex in microcirculation in human cutaneous tissue. Acta Physiol. Scand. **98**(2), 227–231 (1976)

S. Herault, N. Tobal et al., Effect of human head flexion on the control of peripheral blood flow in microgravity and in 1 g. Eur. J. Appl. Physiol. **87**(3), 296–303 (2002)

S.J. Herdman, Treatment of benign paroxysmal positional vertigo. Phys. Ther. **70**(6), 381–388 (1990)

D. Hershkovitz, N. Asna et al., Ondansetron for the prevention of seasickness in susceptible sailors: an evaluation at sea. Aviat. Space Environ. Med. **80**(7), 643–646 (2009)

W. Hildebrandt, H.C. Gunga et al., Enhanced slow caudad fluid shifts in orthostatic intolerance after 24-h bed-rest. Eur. J. Appl. Physiol. Occup. Physiol. **69**(1), 61–70 (1994)

I. Hindmarch, D.B. Fairweather, Assessing the residual effects of hypnotics. Acta Psychiatr. Belg. **94**(2), 88–95 (1994)

I. Hindmarch, S. Johnson et al., The acute and sub-chronic effects of levocetirizine, cetirizine, loratadine, promethazine and placebo on cognitive function, psychomotor performance, and weal and flare. Curr. Med. Res. Opin. **17**(4), 241–255 (2001a)

I. Hindmarch, A. Patat et al., Residual effects of zaleplon and zolpidem following middle of the night administration five hours to one hour before awakening. Hum. Psychopharmacol. **16**(2), 159–167 (2001b)

I. Hindmarch, U. Rigney et al., A naturalistic investigation of the effects of day-long consumption of tea, coffee and water on alertness, sleep onset and sleep quality. Psychopharmacology (Berl.) **149**(3), 203–216 (2000)

R.B. Hoffman, G.A. Salinas et al., Piracetam and fish orientation during parabolic aircraft flight. Aviat. Space Environ. Med. **51**(6), 568–576 (1980)

J. Hollander, M. Gore et al., Spaceflight downregulates antioxidant defense systems in rat liver. Free Radic. Biol. Med. **24**(2), 385–390 (1998)

M.H. Hong, H. Sun et al., Cell-specific activation of the human skeletal alpha-actin by androgens. Endocrinology **149**(3), 1103–1112 (2008)

C.C. Horn, Is there a need to identify new anti-emetic drugs? Drug Discov. Today Ther. Strat. **4**(3), 183–187 (2007)

P.J. Hornby, Central neurocircuitry associated with emesis. Am. J. Med. **111**(Suppl 8A), 106S–112S (2001)

J. Howland, D.J. Rohsenow et al., The effects of transdermal scopolamine on simulated ship navigation and attention/reaction time. Int. J. Occup. Environ. Health **14**(4), 250–256 (2008)

R.E. Hoyt, B.D. Lawson et al., Modafinil as a potential motion sickness countermeasure. Aviat. Space Environ. Med. **80**(8), 709–715 (2009)

S. Hu, W.F. Grant et al., Motion sickness severity and physiological correlates during repeated exposures to a rotating optokinetic drum. Aviat. Space Environ. Med. **62**(4), 308–314 (1991)

J. Huff, F.A. Cucinotta, *Risk of Degenerative Tissue or Other Health Effects from Radiation Exposure* (NASA, Houston TX, HHC, 2008)

P. Hunter, To sleep, perchance to live. Sleeping is vital for health, cognitive function, memory and long life. EMBO Rep. **9**(11), 1070–1073 (2008)

M. Ikuzawa, M. Asashima, Global expression of simulated microgravity-responsive genes in Xenopus liver cells. Zoolog. Sci. **25**(8), 828–837 (2008)

K. Ito, Y. Kagaya et al., Thyroid hormone and chronically unloaded hearts. Vascul. Pharmacol. **52**(3–4), 138–141 (2010)

D.S. Janowsky, S.C. Risch et al., A cholinomimetic model of motion sickness and space adaptation syndrome. Aviat. Space Environ. Med. **55**(8), 692–696 (1984)

R.T. Jennings, J.R. Davis et al., Comparison of aerobic fitness and space motion sickness during the shuttle program. Aviat. Space Environ. Med. **59**(5), 448–451 (1988)

R.T. Jennings, J.P. Stepanek et al., Frequent premature ventricular contractions in an orbital spaceflight participant. Aviat. Space Environ. Med. **81**(6), 597–601 (2010)

B. Jiang, R.R. Roy et al., Absence of a growth hormone effect on rat soleus atrophy during a 4-day spaceflight. J. Appl. Physiol. **74**(2), 527–531 (1993)

Y. Jiang, J.J. Zhao et al., Recombinant human parathyroid hormone (1–34) [teriparatide] improves both cortical and cancellous bone structure. J. Bone Miner. Res. **18**(11), 1932–1941 (2003)

S.H. Jo, H.K. Hong et al., H(1) antihistamine drug promethazine directly blocks hERG K(+) channel. Pharmacol. Res. **60**(5), 429–437 (2009)

K.L. Johansen, K. Mulligan et al., Anabolic effects of nandrolone decanoate in patients receiving dialysis: a randomized controlled trial. JAMA **281**(14), 1275–1281 (1999)

L.B. Johansen, C. Gharib et al., Haematocrit, plasma volume and noradrenaline in humans during simulated weightlessness for 42 days. Clin. Physiol. **17**(2), 203–210 (1997)

M.A. Juergensmeyer, E.A. Juergensmeyer et al., Long-term exposure to spaceflight conditions affects bacterial response to antibiotics. Microgr. Sci. Technol. **12**(1), 41–47 (1999)

M.A. Kacena, P. Todd, Gentamicin: effect on *E. coli* in space. Microgr. Sci. Technol. **XII**(3–4), 135–137 (2000)

K. Kakemi, H. Sezaki et al., Absorption and excretion of drugs. XXXVII. Effect of Ca^{2+} on the absorption of tetracycline from the small intestine. (2). Chem. Pharm. Bull. (Tokyo) **16**(11), 2206–2212 (1968a)

K. Kakemi, H. Sezaki et al., Absorption and excretion of drugs. XXXVI. Effect of Ca^{2+} on the absorption of tetracycline from the small intestine. (1). Chem. Pharm. Bull. (Tokyo) **16**(11), 2200–2205 (1968b)

M. Karbownik, R.J. Reiter, Antioxidative effects of melatonin in protection against cellular damage caused by ionizing radiation. Proc. Soc. Exp. Biol. Med. **225**(1), 9–22 (2000)

A. Karni, D. Tanne et al., Dependence on REM sleep of overnight improvement of a perceptual skill. Science **265**(5172), 679–682 (1994)

M.A. Kass, M. Gordon et al., Compliance with topical timolol treatment. Am. J. Ophthalmol. **103**(2), 188–193 (1987)

R.E. Kates, S.R. Harapat et al., Influence of prolonged recumbency on drug disposition. Clin. Pharmacol. Ther. **28**(5), 624–628 (1980)

M. Kato, B.G. Phillips et al., Effects of sleep deprivation on neural circulatory control. Hypertension **35**(5), 1173–1175 (2000)

B.G. Katzung (ed.), *Basic and Clinical Pharmacology* (McGraw Hill Medical, New York, 2007)

G.A. Keller, M.L. Ponte et al., Other drugs acting on nervous system associated with QT-interval prolongation. Curr. Drug Saf. **5**(1), 105–111 (2010)

T.H. Kelly, R.D. Hienz et al., Crewmember performance before, during, and after spaceflight. J. Exp. Anal. Behav. **84**(2), 227–241 (2005)

A.R. Kennedy, J. Guan et al., Countermeasures against space radiation induced oxidative stress in mice. Radiat. Environ. Biophys. **46**(2), 201–203 (2007)

R.S. Kennedy, A. Graybiel et al., Symptomatology under storm conditions in the North Atlantic in control subjects and in persons with bilateral labyrinthine defects. Acta Otolaryngol. **66**(6), 533–540 (1968)

R.S. Kennedy, R.C. Odenheimer et al., Differential effects of scopolamine and amphetamine on microcomputer-based performance tests. Aviat. Space Environ. Med. **61**(7), 615–621 (1990)

W.D. Killgore, N.L. Grugle et al., Restoration of risk-propensity during sleep deprivation: caffeine, dextroamphetamine, and modafinil. Aviat. Space Environ. Med. **79**(9), 867–874 (2008)

Y. Kitamura, A. Miyoshi et al., Effect of glucocorticoid on upregulation of histamine H1 receptor mRNA in nasal mucosa of rats sensitized by exposure to toluene diisocyanate. Acta Otolaryngol. **124**(9), 1053–1058 (2004)

S. Klosterhalfen, S. Kellermann et al., Latent inhibition of rotation chair-induced nausea in healthy male and female volunteers. Psychosom. Med. **67**(2), 335–340 (2005)

R. Knipling, J. Wang, Revised estimates of the US drowsy driver crash problem size based on general estimates system case reviews. *Thirty-ninth Annual Proceedings of the Association for the Advancement of Automotive Medicine*, (Des Plaines, IL, 1995)

T.B. Knudsen, G.P. Daston, MIAME guidelines. Reprod. Toxicol. **19**(3), 263 (2005)

T. Ko, J. Evenden, The effects of psychotomimetic and putative cognitive-enhancing drugs on the performance of a n-back working memory task in rats. Psychopharmacology (Berl.) **202**(1–3), 67–78 (2009)

R.L. Kohl, Failure of metoclopramide to control emesis or nausea due to stressful angular or linear acceleration. Aviat. Space Environ. Med. **58**(2), 125–131 (1987)

R.L. Kohl, D.S. Calkins et al., Arousal and stability: the effects of five new sympathomimetic drugs suggest a new principle for the prevention of space motion sickness. Aviat. Space Environ. Med. **57**(2), 137–143 (1986)

R.L. Kohl, D.S. Calkins et al., Control of nausea and autonomic dysfunction with terfenadine, a peripherally acting antihistamine. Aviat. Space Environ. Med. **62**(5), 392–396 (1991)

R.L. Kohl, S. MacDonald, New pharmacologic approaches to the prevention of space/motion sickness. J. Clin. Pharmacol. **31**(10), 934–946 (1991)

M. Kojima, K. Kangawa, Ghrelin: structure and function. Physiol. Rev. **85**(2), 495–522 (2005)

K.B. Kopacek, Absorption. The Merck Manual (2007), http://www.merck.com/mmpe/sec20/ch303/ch303c.html. Retrieved 12 Jan 2009

K.B. Kopacek, Bioavailablility. The Merck Manual (2007),, http://www.merck.com/mmpe/sec20/ch303/ch303c.html. Retrieved 12 Jan 2009

I. Kovachevich, S. Kondratenko et al., Pharmacokinetics of acetaminophen adminstered in tablets and capsules under long-term space flight conditions. Pharm. Chem. J. **43**(3), 130–133 (2009)

J. Krasnoff, P. Painter, The physiological consequences of bed rest and inactivity. Adv. Ren. Replace. Ther. **6**(2), 124–132 (1999)

W. Kruse, W. Eggert-Kruse et al., Dosage frequency and drug-compliance behaviour – a comparative study on compliance with a medication to be taken twice or four times daily. Eur. J. Clin. Pharmacol. **41**(6), 589–592 (1991)

W. Kruse, W. Eggert-Kruse et al., Compliance and adverse drug reactions: a prospective study with ethinylestradiol using continuous compliance monitoring. Clin. Investig. **71**(6), 483–487 (1993)

W. Kruse, P. Koch-Gwinner et al., Measurement of drug compliance by continuous electronic monitoring: a pilot study in elderly patients discharged from hospital. J. Am. Geriatr. Soc. **40**(11), 1151–1155 (1992)

W. Kruse, E. Weber, Dynamics of drug regimen compliance – its assessment by microprocessor-based monitoring. Eur. J. Clin. Pharmacol. **38**(6), 561–565 (1990)

E. Kuhn, V. Brodan et al., Metabolic reflection of sleep deprivation. Act Nerv Super (Praha) **11**(3), 165–174 (1969)

J.R. Lackner, A. Graybiel, Head movements in non-terrestrial force environments elicit motion sickness: implications for the etiology of space motion sickness. Aviat. Space Environ. Med. **57**(5), 443–448 (1986)

W.D. Lakin, S.A. Stevens et al., Modeling intracranial pressures in microgravity: the influence of the blood–brain barrier. Aviat. Space Environ. Med. **78**(10), 932–936 (2007)

M.R. Landauer, V. Srinivasan et al., Genistein treatment protects mice from ionizing radiation injury. J. Appl. Toxicol. **23**(6), 379–385 (2003)

A.D. Laposky, J. Bass et al., Sleep and circadian rhythms: key components in the regulation of energy metabolism. FEBS Lett. **582**(1), 142–151 (2008a)

A.D. Laposky, M.A. Bradley et al., Sleep-wake regulation is altered in leptin-resistant (db/db) genetically obese and diabetic mice. Am. J. Physiol. Regul. Integr. Comp. Physiol. **295**(6), R2059–R2066 (2008b)

C.M. Lathers, J.M. Riddle et al., Echocardiograms during six hours of bedrest at head-down and head-up tilt and during space flight. J. Clin. Pharmacol. **33**(6), 535–543 (1993)

C.S. Leach, A review of the consequences of fluid and electrolyte shifts in weightlessness. Acta Astronaut. **6**, 1123–1135 (1979)

C.S. Leach, An overview of the endocrine and metabolic changes in manned space flight. Acta Astronaut. **8**, 977–986 (1981)

C.S. Leach, Metabolism and biochemistry in hypogravity. Acta Astronaut. **23**, 105–108 (1991)

C.S. Leach, W.C. Alexander et al., Adrenal and pituitary response of the Apollo 15 crew members. J. Clin. Endocrinol. Metab. **35**(5), 642–645 (1972)

C.S. Leach, C.P. Alfrey et al., Regulation of body fluid compartments during short-term space-flight. J. Appl. Physiol. **81**(1), 105–116 (1996)

C.S. Leach, S.I. Altchuler et al., The endocrine and metabolic responses to space flight. Med. Sci. Sports Exerc. **15**(5), 432–440 (1983)

C.S. Leach, N.M. Cintron et al., Metabolic changes observed in astronauts. J. Clin. Pharmacol. **31**(10), 921–927 (1991a)

C.S. Leach, L.D. Inners et al., Changes in total body water during spaceflight. J. Clin. Pharmacol. **31**(10), 1001–1006 (1991b)

C.S. Leach, P.C. Johnson et al., The endocrine system in space flight. Acta Astronaut. **17**(2), 161–166 (1988)

A. LeBlanc, V. Schneider et al., Bone mineral and lean tissue loss after long duration space flight. J. Musculoskelet. Neuronal Interact. **1**(2), 157–160 (2000)

A.D. LeBlanc, T.B. Driscol et al., Alendronate as an effective countermeasure to disuse induced bone loss. J. Musculoskelet. Neuronal Interact. **2**(4), 335–343 (2002)

A.D. LeBlanc, E.R. Spector et al., Skeletal responses to space flight and the bed rest analog: a review. J. Musculoskelet. Neuronal Interact. **7**(1), 33–47 (2007)

K.C. Lee, J.D. Ma et al., Pharmacogenomics: bridging the gap between science and practice. J. Am. Pharm. Assoc. **50**(1), e1–e14 (2003). quiz e15-17

B.E. Lehnert, R. Iyer, Exposure to low-level chemicals and ionizing radiation: reactive oxygen species and cellular pathways. Hum. Exp. Toxicol. **21**(2), 65–69 (2002)

S.E. Leucuta, L. Vlase, Pharmacokinetics and metabolic drug interactions. Curr. Clin. Pharmacol. **1**(1), 5–20 (2006)

B.D. Levine, J.H. Zuckerman et al., Cardiac atrophy after bed-rest deconditioning: a nonneural mechanism for orthostatic intolerance. Circulation **96**(2), 517–525 (1997)

M.E. Levine, J.C. Chillas, et al., The effects of serotonin (5-HT3) receptor antagonists on gastric tachyarrhythmia and the symptoms of motion sickness (2000,) Aviat Space Environ Med (2000), http://www.ncbi.nlm.nih.gov/entrez/query.fcgi?cmd=Retrieve&db=PubMed&dopt=Citation&list_uids=11086664 Retrieved 11 Nov 2000.

A.J. Lewy, V.K. Bauer et al., Capturing the circadian rhythms of free-running blind people with 0.5 mg melatonin. Brain Res. **918**(1–2), 96–100 (2001)

J.J. Li, J.C. Sutton et al., Discovery of potent and muscle selective androgen receptor modulators through scaffold modifications. J. Med. Chem. **50**(13), 3015–3025 (2007)

U.A. Liberman, S.R. Weiss et al., Effect of oral alendronate on bone mineral density and the incidence of fractures in postmenopausal osteoporosis. The Alendronate Phase III Osteoporosis Treatment Study Group. N. Engl. J. Med. **333**(22), 1437–1443 (1995)

E. Lilly, Forteo (2004), http://pi.lilly.com/us/forteo-medguide.pdf. p. insert

J. Lim, D.F. Dinges, Sleep deprivation and vigilant attention. Ann. N. Y. Acad. Sci. **1129**, 305–322 (2008)

J. Lim, J.C. Tan et al., Sleep deprivation impairs object-selective attention: a view from the ventral visual cortex. PLoS One **5**(2), e9087 (2010)

K.J. Lindsay, P.J. Coates et al., The genetic basis of tissue responses to ionizing radiation. Br. J. Radiol. **80**(1), S2–S6 (2007)

J.C. Lovejoy, S.R. Smith et al., Low-dose T(3) improves the bed rest model of simulated weightlessness in men and women. Am. J. Physiol. **277**(2 Pt 1), E370–E379 (1999)

S.K. Lu, S. Bai et al., Altered cytochrome P450 and P-glycoprotein levels in rats during simulated weightlessness. Aviat. Space Environ. Med. **73**(2), 112–118 (2002)

J.B. Lucot, R.S. Obach et al., The effect of CP-99994 on the responses to provocative motion in the cat. Br. J. Pharmacol. **120**(1), 116–120 (1997)

L. Macho, M. Fickova et al., Plasma insulin levels and insulin receptors in liver and adipose tissue of rats after space flight. Physiologist **34**(1 Suppl), S90–S91 (1991a)

L. Macho, J. Koska et al., The response of endocrine system to stress loads during space flight in human subject. Adv. Space Res. **31**(6), 1605–1610 (2003)

L. Macho, R. Kvetnansky et al., Effects of exposure to space flight on endocrine regulations in experimental animals. Endocr. Regul. **35**(2), 101–114 (2001)

L. Macho, R. Kvetnansky et al., Effects of space flight on endocrine system function in experimental animals. Environ. Med. **40**(2), 95–111 (1996)

L. Macho, R. Kvetnansky et al., Effect of space flights on plasma hormone levels in man and in experimental animal. Acta Astronaut. **23**, 117–121 (1991b)

L. Macho, S. Nemeth et al., Metabolic changes in the animals subjected to space flight. Acta Astronaut. **9**(6–7), 385–389 (1982)

M. Mackiewicz, K.R. Shockley et al., Macromolecule biosynthesis: a key function of sleep. Physiol. Genomics **31**(3), 441–457 (2007)

C. Madeddu, G. Mantovani, An update on promising agents for the treatment of cancer cachexia. Curr. Opin. Support. Palliat. Care **3**(4), 258–262 (2009)

L. Maggi, L. Segale et al., Chemical and physical stability of hydroxypropylmethylcellulose matrices containing diltiazem hydrochloride after gamma irradiation. J. Pharm. Sci. **92**(1), 131–141 (2003)

L. Maggi, L. Segale et al., Polymers-gamma ray interaction. Effects of gamma irradiation on modified release drug delivery systems for oral administration. Int. J. Pharm. **269**(2), 343–351 (2004)

A. Maillet, G. Gauquelin et al., Blood volume regulating hormones response during two space related simulation protocols: four-week confinement and head-down bed-rest. Acta Astronaut. **35**(8), 547–552 (1995)

A.P. Makris, C.R. Rush et al., Behavioral and subjective effects of d-amphetamine and modafinil in healthy adults. Exp. Clin. Psychopharmacol. **15**(2), 123–133 (2007)

K. Manda, A.L. Bhatia, Pre-administration of beta-carotene protects tissue glutathione and lipid peroxidation status following exposure to gamma radiation. J. Environ. Biol. **24**(4), 369–372 (2003)

K. Manda, M. Ueno et al., Melatonin mitigates oxidative damage and apoptosis in mouse cerebellum induced by high-LET 56Fe particle irradiation. J. Pineal Res. **44**(2), 189–196 (2008a)

K. Manda, M. Ueno et al., Memory impairment, oxidative damage and apoptosis induced by space radiation: ameliorative potential of alpha-lipoic acid. Behav. Brain Res. **187**(2), 387–395 (2008b)

K. Manda, M. Ueno et al., Space radiation-induced inhibition of neurogenesis in the hippocampal dentate gyrus and memory impairment in mice: ameliorative potential of the melatonin metabolite, AFMK. J. Pineal Res. **45**(4), 430–438 (2008c)

K. Manda, M. Ueno et al., Cranial irradiation-induced inhibition of neurogenesis in hippocampal dentate gyrus of adult mice: attenuation by melatonin pretreatment. J. Pineal Res. **46**(1), 71–78 (2009)

K. Manda, M. Ueno et al., alpha-Lipoic acid attenuates x-irradiation-induced oxidative stress in mice. Cell Biol. Toxicol. **23**(2), 129–137 (2007)

T. Mano, Autonomic neural functions in space. Curr. Pharm. Biotechnol. **6**(4), 319–324 (2005)

T. Mano, S. Iwase, Sympathetic nerve activity in hypotension and orthostatic intolerance. Acta Physiol. Scand. **177**(3), 359–365 (2003)

D.S. Martin, J.V. Meck, Presyncopal/non-presyncopal outcomes of post spaceflight stand tests are consistent from flight to flight. Aviat. Space Environ. Med. **75**(1), 65–67 (2004)

E.I. Matsnev, D. Bodo, Experimental assessment of selected antimotion drugs. Aviat. Space Environ. Med. **55**(4), 281–286 (1984)

H. Mattie, W.A. Craig et al., Determinants of efficacy and toxicity of aminoglycosides. J. Antimicrob. Chemother. **24**(3), 281–293 (1989)

M.R. McClung, E.M. Lewiecki et al., Denosumab in postmenopausal women with low bone mineral density. N. Engl. J. Med. **354**(8), 821–831 (2006)

A.J. McLachlan, I. Ramzan, Meals and medicines. Austral.. Prescrib. **29**(2), 40–42 (2006)

J.V. Meck, S. Dreyer et al., Long-duration head-down bed rest: project overview, vital signs, and fluid balance. Aviat. Space Environ. Med. **80**(5, Suppl), A1–A8 (2009)

J.V. Meck, W.W. Waters et al., Mechanisms of postspaceflight orthostatic hypotension: low alpha1-adrenergic receptor responses before flight and central autonomic dysregulation postflight. Am. J. Physiol. Heart Circ. Physiol. **286**(4), H1486–H1495 (2004)

J.P. Meehan, Biosatellite 3: a physiological interpretation. Life Sci. Space Res. **9**, 83–98 (1971)

D. Megighian, A. Martini, Motion sickness and space sickness: clinical and experimental findings. ORL J. Otorhinolaryngol. Relat. Spec. **42**(4), 185–195 (1980)

S.K. Mehta, R.J. Cohrs et al., Stress-induced subclinical reactivation of varicella zoster virus in astronauts. J. Med. Virol. **72**(1), 174–179 (2004)

H.K. Meier-Ewert, P.M. Ridker et al., Effect of sleep loss on C-reactive protein, an inflammatory marker of cardiovascular risk. J. Am. Coll. Cardiol. **43**(4), 678–683 (2004)

A.H. Merrill Jr., M. Hoel et al., Altered carbohydrate, lipid, and xenobiotic metabolism by liver from rats flown on Cosmos 1887. FASEB J. **4**(1), 95–100 (1990)

A.H. Merrill Jr., E. Wang et al., Hepatic function in rats after spaceflight: effects on lipids, glycogen, and enzymes. Am. J. Physiol. **252**(2 Pt 2), R222–R226 (1987)

A.H. Merrill Jr., E. Wang et al., Differences in glycogen, lipids, and enzymes in livers from rats flown on COSMOS 2044. J. Appl. Physiol. **73**(2 Suppl), 142S–147S (1992)

M. Mieda, T. Sakurai, Integrative physiology of orexins and orexin receptors. CNS Neurol. Disord. Drug Targets **8**(4), 281–295 (2009)

P.F. Migeotte, G.K. Prisk et al., Microgravity alters respiratory sinus arrhythmia and short-term heart rate variability in humans. Am. J. Physiol. Heart Circ. Physiol. **284**(6), H1995–H2006 (2003)

J.N. Miner, W. Chang et al., An orally active selective androgen receptor modulator is efficacious on bone, muscle, and sex function with reduced impact on prostate. Endocrinology **148**(1), 363–373 (2007)

T. Misaki, N. Kyoda et al., Timing and side effects of flumazenil for dental outpatients receiving intravenous sedation with midazolam. Anesth. Prog. **44**(4), 127–131 (1997)

M. Miyamoto, Pharmacology of ramelteon, a selective MT1/MT2 receptor agonist: a novel therapeutic drug for sleep disorders. CNS Neurosci. Ther. **15**(1), 32–51 (2009)

D.J. Mollicone, H.P. Van Dongen et al., Response surface mapping of neurobehavioral performance: testing the feasibility of split sleep schedules for space operations. Acta Astronaut. **63**(7–10), 833–840 (2008)

T.H. Monk, D.J. Buysse et al., Sleep and circadian rhythms in four orbiting astronauts. J. Biol. Rhythms **13**(3), 188–201 (1998)

T.H. Monk, K.S. Kennedy et al., Decreased human circadian pacemaker influence after 100 days in space: a case study. Psychosom. Med. **63**(6), 881–885 (2001)

L.D. Montgomery, Body volume changes during simulated microgravity I: technique and comparison of men and women during horizontal bed rest. Aviat. Space Environ. Med. **64**(10), 893–898 (1993)

L.D. Montgomery, A.J. Parmet et al., Body volume changes during simulated microgravity: auditory changes, segmental fluid redistribution, and regional hemodynamics. Ann. Biomed. Eng. **21**(4), 417–433 (1993)

J.E. Morley, F.E. Kaiser et al., Longitudinal changes in testosterone, luteinizing hormone, and follicle-stimulating hormone in healthy older men. Metabolism **46**(4), 410–413 (1997)

G.R. Morrow, Susceptibility to motion sickness and chemotherapy-induced side-effects. Lancet **1**(8373), 390–391 (1984)

P.J. Muller, J. Vernikos-Danellis, Alteration in drug toxicity by environmental variables. Proc. West Pharmacol. Soc. **11**, 52–53 (1968)

J.M. Mullington, M. Haack et al., Cardiovascular, inflammatory, and metabolic consequences of sleep deprivation. Prog. Cardiovasc. Dis. **51**(4), 294–302 (2009)

S.L. Mulvagh, J.B. Charles et al., Echocardiographic evaluation of the cardiovascular effects of short-duration spaceflight. J. Clin. Pharmacol. **31**(10), 1024–1026 (1991)

J.S. Murley, Y. Kataoka et al., Delayed radioprotection by nuclear transcription factor kappaB-mediated induction of manganese superoxide dismutase in human microvascular endothelial cells after exposure to the free radical scavenger WR1065. Free Radic. Biol. Med. **40**(6), 1004–1016 (2006)

E.R. Muth, Motion and space sickness: intestinal and autonomic correlates. Auton. Neurosci. **129**(1–2), 58–66 (2006)

Z. Nachum, A. Shupak et al., Transdermal scopolamine for prevention of motion sickness: clinical pharmacokinetics and therapeutic applications. Clin. Pharmacokinet. **45**(6), 543–566 (2006)

R. Narayanan, M.L. Mohler et al., Selective androgen receptor modulators in preclinical and clinical development. Nucl. Recept. Signal. **6**, e010 (2008)

R.M. Neer, C.D. Arnaud et al., Effect of parathyroid hormone (1–34) on fractures and bone mineral density in postmenopausal women with osteoporosis. N. Engl. J. Med. **344**(19), 1434–1441 (2001)

A.N. Nicholson, Sleep patterns in the aerospace environment. Proc. R. Soc. Med. **65**(2), 192–194 (1972)

C.A. Nickerson, C.M. Ott et al., Microgravity as a novel environmental signal affecting *Salmonella enterica* serovar Typhimurium virulence. Infect. Immun. **68**(6), 3147–3152 (2000)

C.A. Nickerson, C.M. Ott et al., Low-shear modeled microgravity: a global environmental regulatory signal affecting bacterial gene expression, physiology, and pathogenesis. J. Microbiol. Methods **54**(1), 1–11 (2003)

C.A. Nickerson, C.M. Ott et al., Microbial responses to microgravity and other low-shear environments. Microbiol. Mol. Biol. Rev. **68**(2), 345–361 (2004)

NIH, NIH renews major research program to develop medical countermeasures against radiological and nuclear threats (2010), http://www.nih.gov/news/health/aug2010/niaid-19.htm

W.S. Nimmo, L.F. Prescott, The influence of posture on paracetemol absorption. Br. J. Clin. Pharmacol. **5**, 348–349 (1978)

W.T. Norfleet, J.J. Degioanni et al., Treatment of motion sickness in parabolic flight with buccal scopolamine. Aviat. Space Environ. Med. **63**(1), 46–51 (1992)

P. Norsk, N.J. Christensen et al., Unexpected renal responses in space. Lancet **356**(9241), 1577–1578 (2000)

P. Norsk, C. Drummer et al., Renal and endocrine responses in humans to isotonic saline infusion during microgravity. J. Appl. Physiol. **78**, 2253–2259 (1995)

M.C. Noverr, G.B. Huffnagle, Does the microbiota regulate immune responses outside the gut? Trends Microbiol. **12**(12), 562–568 (2004)

A. Nunez, M.L. Rodrigo-Angulo et al., Hypocretin/orexin neuropeptides: participation in the control of sleep-wakefulness cycle and energy homeostasis. Curr. Neuropharmacol. **7**(1), 50–59 (2009)

C.V. Odvina, S. Levy et al., Unusual mid-shaft fractures during long-term bisphosphonate therapy. Clin. Endocrinol. (Oxf.) **72**(2), 161–168 (2010)

Y. Ogawa, T. Kanbayashi et al., Total sleep deprivation elevates blood pressure through arterial baroreflex resetting: a study with microneurographic technique. Sleep **26**(8), 986–989 (2003)

I.V. Ogneva, V.A. Kurushin et al., Effect of short-term gravitational unloading on rat and mongolian gerbil muscles. J. Muscle Res. Cell Motil. **30**(7–8), 261–265 (2010)

C.M. Oman, Motion sickness: a synthesis and evaluation of the sensory conflict theory. Can. J. Physiol. Pharmacol. **68**(2), 294–303 (1990)

C.M. Oman, Sensory conflict theory and space sickness: our changing perspective. J. Vestib. Res. **8**(1), 51–56 (1998)

B. Oosterhuis, J.H. Jonkman, Pharmacokinetic studies in healthy volunteers in the context of in vitro/in vivo correlations. Eur. J. Drug Metab. Pharmacokinet. **18**(1), 19–30 (1993)

J.H.J. Ortega, D.L. Harm, Space and entry motion sickness, in *Principles of Clinical Medicine for Space Flight*, ed. by M.R. Barrat, S.L. Pool (Springer, New York, 2008)

I. Oswald, The function of sleep. Postgrad. Med. J. **52**(603), 15–18 (1976)

S. Otmani, A. Demazieres et al., Effects of prolonged-release melatonin, zolpidem, and their combination on psychomotor functions, memory recall, and driving skills in healthy middle aged and elderly volunteers. Hum. Psychopharmacol. **23**(8), 693–705 (2008)

S.T. Page, B.T. Marck et al., Tissue selectivity of the anabolic steroid, 19-nor-4-androstenediol-3beta,17beta-diol in male Sprague Dawley rats: selective stimulation of muscle mass and bone mineral density relative to prostate mass. Endocrinology **149**(4), 1987–1993 (2008)

S.C. Pageau, Denosumab. MAbs **1**(3), 210–215 (2009)

W. Pang, C. Li et al., The environmental light influences the circulatory levels of retinoic acid and associates with hepatic lipid metabolism. Endocrinology **149**(12), 6336–6342 (2008)

D.E. Parker, Labyrinth and cerebral-spinal fluid pressure changes in guinea pigs and monkeys during simulated zero G. Aviat. Space Environ. Med. **48**(4), 356–361 (1977)

L.S. Parnes, S.K. Agrawal et al., Diagnosis and management of benign paroxysmal positional vertigo (BPPV). CMAJ **169**(7), 681–693 (2003)

A.C. Parrott, K. Wesnes, Promethazine, scopolamine and cinnarizine: comparative time course of psychological performance effects. Psychopharmacology (Berl.) **92**(4), 513–519 (1987)

A. Patat, I. Paty et al., Pharmacodynamic profile of Zaleplon, a new non-benzodiazepine hypnotic agent. Hum. Psychopharmacol. **16**(5), 369–392 (2001)

M.G. Paule, J.J. Chelonis et al., Effects of drug countermeasures for space motion sickness on working memory in humans. Neurotoxicol. Teratol. **26**(6), 825–837 (2004)

A. Pavy-Le Traon, A. Guell et al., The use of medicaments in space–therapeutic measures and potential impact of pharmacokinetics due to weightlessness. ESA J **18**(1), 33–50 (1994)

A. Pavy-Le Traon, M. Heer et al., From space to Earth: advances in human physiology from 20 years of bed rest studies (1986–2006). Eur. J. Appl. Physiol. **101**(2), 143–194 (2007)

D.A. Payne, S.K. Mehta et al., Incidence of Epstein-Barr virus in astronaut saliva during spaceflight. Aviat. Space Environ. Med. **70**(12), 1211–1213 (1999)

M.J. Pecaut, D.S. Gridley, The impact of mouse strain on iron ion radio-immune response of leukocyte populations. Int. J. Radiat. Biol. **86**(5), 409–419 (2010)

M.A. Perhonen, F. Franco et al., Cardiac atrophy after bed rest and spaceflight. J. Appl. Physiol. **91**(2), 645–653 (2001)

K. Petrie, A.G. Dawson et al., A double-blind trial of melatonin as a treatment for jet lag in international cabin crew. Biol. Psychiatry **33**(7), 526–530 (1993)

D.L. Pierson, R.P. Stowe et al., Epstein-Barr virus shedding by astronauts during space flight. Brain Behav. Immunol. **19**(3), 235–242 (2005)

R.A. Pietrzyk, J.A. Jones et al., Renal stone formation among astronauts. Aviat. Space Environ. Med. **78**(4 Suppl), A9–A13 (2007)

J.R. Plant, D.B. MacLeod, Response of a promethazine-induced coma to flumazenil. Ann. Emerg. Med. **24**(5), 979–982 (1994)

S. Platts, *Risk of Cardiac Rhythm Problems During Spaceflight* (NASA, Houston TX, HHC, 2008a)

S. Platts, *Risk of Orthostatic Intolerance During Re-exposure to Gravity* (NASA, Houston TX, HHC, 2008b)

S.H. Platts, D.S. Martin et al., Cardiovascular adaptations to long-duration head-down bed rest. Aviat. Space Environ. Med. **80**(5 Suppl), A29–A36 (2009a)

S.H. Platts, S.J. Shi et al., Akathisia with combined use of midodrine and promethazine. JAMA **295**(17), 2000–2001 (2006a)

S.H. Platts, J.A. Tuxhorn et al., Compression garments as countermeasures to orthostatic intolerance. Aviat. Space Environ. Med. **80**(5), 437–442 (2009b)

S.H. Platts, M.G. Ziegler et al., Hemodynamic effects of midodrine after spaceflight in astronauts without orthostatic hypotension. Aviat. Space Environ. Med. **77**(4), 429–433 (2006b)

O. Pompeiano, P. d'Ascanio et al., Gene expression in rat vestibular and reticular structures during and after space flight. Neuroscience **114**(1), 135–155 (2002)

L. Putcha, Pharmacotherapeutics in space. J. Gravit. Physiol. **6**(1), P165–P168 (1999)

L. Putcha, Data mining – Pharmacotherapeutics of space motion sickness (2009)

L. Putcha, K.L. Berens et al., Pharmaceutical use by U.S. astronauts on space shuttle missions. Aviat. Space Environ. Med. **70**(7), 705–708 (1999)

L. Putcha, K.J. Tietze et al., Bioavailability of intranasal scopolamine in normal subjects. J. Pharm. Sci. **85**(8), 899–902 (1996)

C. Queckenberg, U. Fuhr, Influence of posture on pharmacokinetics. Eur. J. Clin. Pharmacol. **65**(2), 109–119 (2009)

C. Queckenberg, J. Meins et al., Absorption, pharmacokinetics and safety of triclosan after dermal administration. Antimicrob. Agents Chemother. **54**(1), 570–572 (2009)

R.N. Racine, S.M. Cormier, Effect of spaceflight on rat hepatocytes: a morphometric study. J. Appl. Physiol. **73**(2 Suppl), 136S–141S (1992)

P.C. Rambaut, C.S. Leach et al., Observations in energy balance in man during spaceflight. Am. J. Physiol. **233**(5), R208–R212 (1977a)

P.C. Rambaut, C.S. Leach et al., Metabolic energy requirements during manned orbital Skylab missions. Life Sci. Space Res. **15**, 187–191 (1977b)

C.D. Ramsdell, T.J. Mullen et al., Midodrine prevents orthostatic intolerance associated with simulated spaceflight. J. Appl. Physiol. **90**(6), 2245–2248 (2001)

C.A. Ray, New insights into orthostatic hypotension. Am. J. Physiol. Regul. Integr. Comp. Physiol. **294**(5), R1575–R1576 (2008)

J.T. Reason, J.J. Brand, *Motion Sickness* (Academic Press, London, 1975)

A. Rechtscheffen, A. Kales (eds.), *A Manual for Standardaized Terminology and Scoring System for Sleep Stages of Human Subjects* (Public Health Service, US Government Printing Office, Washington DC, 1968)

K. Reid, J.L. Palmer et al., Comparison of the neurokinin-1 antagonist GR205171, alone and in combination with the 5-HT3 antagonist ondansetron, hyoscine and placebo in the prevention of motion-induced nausea in man. Br. J. Clin. Pharmacol. **50**(1), 61–64 (2000)

R.J. Reiter, D.X. Tan et al., Melatonin as an antioxidant: physiology versus pharmacology. J. Pineal Res. **39**(2), 215–216 (2005)

A.G. Renwick, C.H. Ahsan et al., The influence of posture on the pharmacokinetics of orally administered nifedipine. Br. J. Clin. Pharmacol. **34**(4), 332–336 (1992)

M.F. Reschke, J.J. Bloomberg et al., Posture, locomotion, spatial orientation, and motion sickness as a function of space flight. Brain Res. Brain Res. Rev. **28**(1–2), 102–117 (1998)

G. Ricci, A. Catizone et al., Microgravity effect on testicular functions. J. Gravit. Physiol. **11**(2), P61–P62 (2004)

G. Ricci, R. Esposito et al., Direct effects of microgravity on testicular function: analysis of hystological, molecular and physiologic parameters. J. Endocrinol. Invest. **31**(3), 229–237 (2008)

L. Rice, W. Ruiz et al., Neocytolysis on descent from altitude: a newly recognized mechanism for the control of red cell mass. Ann. Intern. Med. **134**(8), 652–656 (2001)

G.S. Richardson, P.C. Zee et al., Circadian phase-shifting effects of repeated ramelteon administration in healthy adults. J. Clin. Sleep Med. **4**(5), 456–461 (2008)

F. Ridout, I. Hindmarch, The effects of acute doses of fexofenadine, promethazine, and placebo on cognitive and psychomotor function in healthy Japanese volunteers. Ann. Allergy Asthma Immunol. **90**(4), 404–410 (2003)

D.W. Rimmer, D.B. Boivin et al., Dynamic resetting of the human circadian pacemaker by intermittent bright light. Am. J. Physiol. Regul. Integr. Comp. Physiol. **279**(5), R1574–R1579 (2000)

J. Rittweger, H.M. Frost et al., Muscle atrophy and bone loss after 90 days' bed rest and the effects of flywheel resistive exercise and pamidronate: results from the LTBR study. Bone **36**(6), 1019–1029 (2005)

M.S. Roberts, M.J. Denton, Effect of posture and sleep on pharmacokinetics. I. Amoxycillin. Eur. J. Clin. Pharmacol. **18**(2), 175–183 (1980)

M.D. Ross, D.L. Tomko, Effect of gravity on vestibular neural development. Brain Res. Brain Res. Rev. **28**(1–2), 44–51 (1998)

A.C. Rossum, M.L. Wood et al., Evaluation of cardiac rhythm disturbances during extravehicular activity. Am. J. Cardiol. **79**(8), 1153–1155 (1997)

A.C. Rossum, M.G. Ziegler et al., Effect of spaceflight on cardiovascular responses to upright posture in a 77-year-old astronaut. Am. J. Cardiol. **88**(11), 1335–1337 (2001)

M. Rowland, Hemodynamic factors in pharmacokinetics. Triangle **14**(3–4), 109–116 (1975)

T.A. Roy, M.R. Blackman et al., Interrelationships of serum testosterone and free testosterone index with FFM and strength in aging men. Am. J. Physiol. Endocrinol. Metab. **283**(2), E284–E294 (2002)

R.H. Rumble, M.S. Roberts et al., Effects of posture and sleep on the pharmacokinetics of paracetamol (acetaminophen) and its metabolites. Clin. Pharmacokinet. **20**(2), 167–173 (1991)

R.L. Sack, D. Auckley et al., Circadian rhythm sleep disorders: part I, basic principles, shift work and jet lag disorders. An American Academy of Sleep Medicine review. Sleep **30**(11), 1460–1483 (2007)

R.L. Sack, R.J. Hughes et al., Sleep-promoting effects of melatonin: at what dose, in whom, under what conditions, and by what mechanisms? Sleep **20**(10), 908–915 (1997)

S.D. Saini, P. Schoenfeld et al., Effect of medication dosing frequency on adherence in chronic diseases. Am. J. Manag. Care **15**(6), e22–e33 (2009)

S. Saivin, A. Pavy-Le Traon et al., Impact of a four-day head-down tilt (−6 degrees) on lidocaine pharmacokinetics used as probe to evaluate hepatic blood flow. J. Clin. Pharmacol. **35**(7), 697–704 (1995)

S. Saivin, A. Pavy-Le Traon et al., Pharmacology in space: pharmacokinetics. Adv. Space Biol. Med. **6**, 107–121 (1997)

G.J. Sanger, P.L. Andrews, Treatment of nausea and vomiting: gaps in our knowledge. Auton. Neurosci. **129**(1–2), 3–16 (2006)

D. Santucci, N. Francia et al., A mouse model of neurobehavioural response to altered gravity conditions: an ontogenetical study. Behav. Brain Res. **197**(1), 109–118 (2009)

P.A. Santy, M.W. Bungo, Pharmacologic considerations for shuttle astronauts. J. Clin. Pharmacol. **31**(10), 931–933 (1991)

G. Sato, A. Uno et al., Effects of hypergravity on histamine H1 receptor mRNA expression in hypothalamus and brainstem of rats: implications for development of motion sickness. Acta Otolaryngol. **129**(1), 45–51 (2009)

I. Sayet, G. Neuilly et al., Influence of spaceflight, hindlimb suspension, and venous occlusion on alpha 1-adrenoceptors in rat vena cava. J. Appl. Physiol. **78**(5), 1882–1888 (1995)

M.B. Scharf, T. Roth et al., A multicenter, placebo-controlled study evaluating zolpidem in the treatment of chronic insomnia. J. Clin. Psychiatry **55**(5), 192–199 (1994)

P. Schlyter, Radiometry and photometry in astronmoy FAQ (2006), Retrieved 11/12/2009, from, http://stjarnhimlen.se/comp/radfaq.html#10

B.O. Schneeman, Gastrointestinal physiology and functions. Br. J. Nutr. **88**(Suppl 2), S159–S163 (2002)

S. Schneider, V. Brummer et al., Parabolic flight experience is related to increased release of stress hormones. Eur. J. Appl. Physiol. **100**(3), 301–308 (2007)

D.J. Schroeder, W.E. Collins et al., Effects of some motion sickness suppressants on static and dynamic tracking performance. Aviat. Space Environ. Med. **56**(4), 344–350 (1985)

E.L. Schuck, M. Grant et al., Effect of simulated microgravity on the disposition and tissue penetration of ciprofloxacin in healthy volunteers. J. Clin. Pharmacol. **45**(7), 822–831 (2005)

S. Schutte-Rodin, L. Broch et al., Clinical guideline for the evaluation and management of chronic insomnia in adults. J. Clin. Sleep Med. **4**(5), 487–504 (2008)

H.S. Sedlacek, D.S. Ramsey et al., Comparative efficacy of maropitant and selected drugs in preventing emesis induced by centrally or peripherally acting emetogens in dogs. J. Vet. Pharmacol. Ther. **31**(6), 533–537 (2008)

D.E. Sellmeyer, M. Schloetter et al., Potassium citrate prevents increased urine calcium excretion and bone resorption induced by a high sodium chloride diet. J. Clin. Endocrinol. Metab. **87**(5), 2008–2012 (2002)

G. Sener, L. Kabasakal et al., Ginkgo biloba extract protects against ionizing radiation-induced oxidative organ damage in rats. Pharmacol. Res. **53**(3), 241–252 (2006)

H. Senzaki, T. Yasui et al., Alendronate inhibits urinary calcium microlith formation in a three-dimensional culture model. Urol. Res. **32**(3), 223–228 (2004)

J.M. Serrador, T.T. Schlegel et al., Cerebral hypoperfusion precedes nausea during centrifugation. Aviat. Space Environ. Med. **76**(2), 91–96 (2005)

J.M. Serrador, T.T. Schlegel et al., Vestibular effects on cerebral blood flow. BMC Neurosci. **10**, 119 (2009)

J.M. Serrador, J.K. Shoemaker et al., Cerebral vasoconstriction precedes orthostatic intolerance after parabolic flight. Brain Res. Bull. **53**(1), 113–120 (2000)

K.G. Shah, R. Wu et al., Human ghrelin ameliorates organ injury and improves survival after radiation injury combined with severe sepsis. Mol. Med. **15**(11–12), 407–414 (2009)

L. Shargel, S. Wu-Pong et al., *Applied Biopharmaceutics and Pharmacokinetics* (McGraw Hill, New York, 2005)

S.-J. Shi, S.H. Platts, et al.,Effects of midodrine, promethazine, and their combination on orthostatic intolerance in normal subjects. Aviat. Space Environ. Med., in review (2010)

S.J. Shi, D.A. South et al., Fludrocortisone does not prevent orthostatic hypotension in astronauts after spaceflight. Aviat. Space Environ. Med. **75**(3), 235–239 (2004)

V.M. Shilov, N.N. Lizko et al., Changes in the microflora of man during long-term confinement. Life Sci. Space Res. **9**, 43–49 (1971)

M. Shiraishi, M. Schou et al., Comparison of acute cardiovascular responses to water immersion and head-down tilt in humans. J. Appl. Physiol. **92**(1), 264–268 (2002)

J. Sibonga, *Risk of Accelerated Osteoporosis* (NASA, Houston TX, HHC, 2008a)

J. Sibonga, *Risk of Bone Fracture* (NASA, Houston TX, HHC, 2008b)

J.M. Siegel, Sleep viewed as a state of adaptive inactivity. Nat. Rev. Neurosci. **10**(10), 747–753 (2009)

R. Sih, J.E. Morley et al., Testosterone replacement in older hypogonadal men: a 12-month randomized controlled trial. J. Clin. Endocrinol. Metab. **82**(6), 1661–1667 (1997)

N. Simpson, D.F. Dinges, Sleep and inflammation. Nutr. Rev. **65**(12 Pt 2), S244–S252 (2007)

A.N. Siriwardena, M.Z. Qureshi et al., Magic bullets for insomnia? Patients' use and experiences of newer (Z drugs) versus older (benzodiazepine) hypnotics for sleep problems in primary care. Br. J. Gen. Pract. **58**(551), 417–422 (2008)

A.N. Siriwardena, Z. Qureshi et al., GPs' attitudes to benzodiazepine and 'Z-drug' prescribing: a barrier to implementation of evidence and guidance on hypnotics. Br. J. Gen. Pract. **56**(533), 964–967 (2006)

J.T. Slattery, J.M. Wilson et al., Dose-dependent pharmacokinetics of acetaminophen: evidence of glutathione depletion in humans. Clin. Pharmacol. Ther. **41**(4), 413–418 (1987)

R.F. Smith, K. Stanton et al., Quantitative electrocardiography during extended space flight: the second manned Skylab mission. Aviat. Space Environ. Med. **47**(4), 353–359 (1976)

S.M. Smith, M.E. Wastney et al., Calcium metabolism before, during, and after a 3-mo spaceflight: kinetic and biochemical changes. Am. J. Physiol. **277**(1 Pt 2), R1–R10 (1999)

S.M. Smith, S.R. Zwart, Nutritional biochemistry of spaceflight. Adv. Clin. Chem. **46**, 87–130 (2008)

S.M. Smith, S.R. Zwart et al., Nutrient-drug interactions, in *Nutritional Biochemistry of Space Flight* (Nova Science, New York, 2009a)

S.M. Smith, S.R. Zwart et al., Space programs and space food systems, in *Nutritional Biochemistry of Space Flight* (Nova Science, New York, 2009b)

P.J. Snyder, H. Peachey et al., Effects of testosterone replacement in hypogonadal men. J. Clin. Endocrinol. Metab. **85**(8), 2670–2677 (2000)

M.D. Sockol, D.A. Raichlen et al., Chimpanzee locomotor energetics and the origin of human bipedalism. Proc. Natl. Acad. Sci. USA **104**(30), 12265–12269 (2007)

Q.H. Song, K. Toriizuka et al., Effect of Kampo herbal medicines on murine water metabolism in a microgravity environment. Am. J. Chin. Med. **30**(4), 617–627 (2002)

D. Soyal, A. Jindal et al., Modulation of radiation-induced biochemical alterations in mice by rosemary (*Rosemarinus officinalis*) extract. Phytomedicine **14**(10), 701–705 (2007)

E.R. Spector, S.M. Smith et al., Skeletal effects of long-duration head-down bed rest. Aviat. Space Environ. Med. **80**(5 Suppl), A23–A28 (2009)

H.A. Spiller, D. Borys et al., Toxicity from modafinil ingestion. Clin. Toxicol. (Phila.) **47**(2), 153–156 (2009)

V. Srinivasan, S.R. Pandi-Perumal et al., Melatonin and melatonergic drugs on sleep: possible mechanisms of action. Int. J. Neurosci. **119**(6), 821–846 (2009)

V. Srinivasan, D.W. Spence et al., Jet lag: therapeutic use of melatonin and possible application of melatonin analogs. Travel Med. Infect. Dis. **6**(1–2), 17–28 (2008)

T.P. Stein, M.J. Leskiw, Oxidant damage during and after spaceflight. Am. J. Physiol. Endocrinol. Metab. **278**(3), E375–E382 (2000)

P.C. Stepaniak, S.R. Ramchandani et al., Acute urinary retention among astronauts. Aviat. Space Environ. Med. **78**(4 Suppl), A5–A8 (2007)

R.M. Stern, K.L. Koch et al., Tachygastria and motion sickness. Aviat. Space Environ. Med. **56**(11), 1074–1077 (1985)

D.R. Stickney, C. Dowding et al., 5-androstenediol improves survival in clinically unsupported rhesus monkeys with radiation-induced myelosuppression. Int. Immunopharmacol. **7**(4), 500–505 (2007)

J.R. Stott, G.R. Barnes et al., The effect on motion sickness and oculomotor function of GR 38032F, a 5-HT3-receptor antagonist with anti-emetic properties. Br. J. Clin. Pharmacol. **27**(2), 147–157 (1989)

A. Sullivan, C. Edlund et al., Effect of antimicrobial agents on the ecological balance of human microflora. Lancet Infect. Dis. **1**(2), 101–114 (2001)

W.K. Sumanasekera, G.U. Sumanasekera et al., Estradiol and dihydrotestosterone regulate endothelial cell barrier function after hypergravity-induced alterations in MAPK activity. Am. J. Physiol. Cell Physiol. **293**(2), C566–C573 (2007)

R.L. Summers, D.S. Martin et al., Mechanism of spaceflight-induced changes in left ventricular mass. Am. J. Cardiol. **95**(9), 1128–1130 (2005)

N.R. Swerdlow, M.A. Geyer, Using an animal model of deficient sensorimotor gating to study the pathophysiology and new treatments of schizophrenia. Schizophr. Bull. **24**(2), 285–301 (1998)

T. Taddeo, C. Armstrong, Spaceflight medical systems, in *Principles of Clinical Medicine for Space Flight*, ed. by M. Barratt, S. Pool (Springer, New York, 2008)

B. Takacs, D. Hanak, A prototype home robot with an ambient facial interface to improve drug compliance. J. Telemed. Telecare **14**(7), 393–395 (2008)

R. Tamma, G. Colaianni et al., Oxytocin is an anabolic bone hormone. Proc. Natl. Acad. Sci. USA **106**(17), 7149–7154 (2009)

D.X. Tan, L.C. Manchester et al., One molecule, many derivatives: a never-ending interaction of melatonin with reactive oxygen and nitrogen species? J. Pineal Res. **42**(1), 28–42 (2007)

D.D. Taub, Novel connections between the neuroendocrine and immune systems: the ghrelin immunoregulatory network. Vitam. Horm. **77**, 325–346 (2008)

K.H. Taylor, L.S. Middlefell et al., Osteonecrosis of the jaws induced by anti-RANK ligand therapy. Br. J. Oral Maxillofac. Surg. **48**(3), 221–223 (2009)

D.G. Thompson, A.S. Mason et al., Mineralocorticoid replacement in Addison's disease. Clin. Endocrinol. (Oxf.) **10**(5), 499–506 (1979)

W.E. Thornton, T.P. Moore et al., Clinical characterization and etiology of space motion sickness. Aviat. Space Environ. Med. **58**(9 Pt 2), A1–A8 (1987)

M.E. Tischler, E.J. Henriksen et al., Spaceflight on STS-48 and earth-based unweighting produce similar effects on skeletal muscle of young rats. J. Appl. Physiol. **74**(5), 2161–2165 (1993)

R. Tixador, G. Richoilley et al., Study of minimal inhibitory concentration of antibiotics on bacteria cultivated in vitro in space (Cytos 2 experiment). Aviat. Space Environ. Med. **56**(8), 748–751 (1985)

B.W. Tobin, P.N. Uchakin et al., Insulin secretion and sensitivity in space flight: diabetogenic effects. Nutrition **18**(10), 842–848 (2002)

O. Tochikubo, A. Ikeda et al., Effects of insufficient sleep on blood pressure monitored by a new multibiomedical recorder. Hypertension **27**(6), 1318–1324 (1996)

N. Treister, N. Sheehy et al., Dental panoramic radiographic evaluation in bisphosphonate-associated osteonecrosis of the jaws. Oral Dis. **15**(1), 88–92 (2009)

D. Tricarico, A. Mele et al., Phenotype-dependent functional and pharmacological properties of BK channels in skeletal muscle: effects of microgravity. Neurobiol. Dis. **20**(2), 296–302 (2005)

R. Tricker, R. Casaburi et al., The effects of supraphysiological doses of testosterone on angry behavior in healthy eugonadal men – a clinical research center study. J. Clin. Endocrinol. Metab. **81**(10), 3754–3758 (1996)

K. Tsuji, P.D. Rahn et al., ^{60}Co-irradiation as an alternate method for sterilization of penicillin G, neomycin, novobiocin, and dihydrostreptomycin. J. Pharm. Sci. **72**(1), 23–26 (1983)

P.L. Turner, M.A. Mainster, Circadian photoreception: ageing and the eye's important role in systemic health. Br. J. Ophthalmol. **92**(11), 1439–1444 (2008)

K. Turnheim, When drug therapy gets old: pharmacokinetics and pharmacodynamics in the elderly. Exp. Gerontol. **38**(8), 843–853 (2003)

K. Turnheim, Drug therapy in the elderly. Exp. Gerontol. **39**(11–12), 1731–1738 (2004)

J. Urquhart, B. Vrijens, 'Hedged' prescribing for partially compliant patients. Clin. Pharmacokinet. **45**(1), 105–108 (2006)

H.P. Van Dongen, D.F. Dinges, Sleep, circadian rhythms, and psychomotor vigilance. Clin. Sports Med. **24**(2), 237–249 (2005). vii-viii

H.P. Van Dongen, N.J. Price et al., Caffeine eliminates psychomotor vigilance deficits from sleep inertia. Sleep **24**(7), 813–819 (2001)

A. van Oeveren, M. Motamedi et al., Discovery of 6-N, N-bis(2,2,2-trifluoroethyl)amino-4-trifluoromethylquinolin-2(1H)-one as a novel selective androgen receptor modulator. J. Med. Chem. **49**(21), 6143–6146 (2006)

C.G. Vecsey, G.S. Baillie et al., Sleep deprivation impairs cAMP signalling in the hippocampus. Nature **461**(7267), 1122–1125 (2009)

B. Verheyden, J. Liu et al., Adaptation of heart rate and blood pressure to short and long duration space missions. Respir. Physiol. Neurobiol. **169**(Suppl 1), S13–S16 (2009)

J. Vernikos, V.A. Convertino, Advantages and disadvantages of fludrocortisone or saline load in preventing post-spaceflight orthostatic hypotension. Acta Astronaut. **33**, 259–266 (1994)

J. Vernikos, M.F. Dallman et al., Drug effects on orthostatic intolerance induced by bedrest. J. Clin. Pharmacol. **31**(10), 974–984 (1991)

B. Vrijens, J. Urquhart, Patient adherence to prescribed antimicrobial drug dosing regimens. J. Antimicrob. Chemother. **55**(5), 616–627 (2005)

C.E. Wade, M.M. Miller et al., Body mass, energy intake, and water consumption of rats and humans during space flight. Nutrition **18**(10), 829–836 (2002)

C.E. Wade, T.J. Wang et al., Rat growth, body composition, and renal function during 30 days increased ambient CO_2 exposure. Aviat. Space Environ. Med. **71**(6), 599–609 (2000)

J.K. Walsh, C.L. Engelhardt, Trends in the pharmacologic treatment of insomnia. J. Clin. Psychiatry **53**, 10–17 (1992). discussion 18

J.K. Walsh, G.W. Vogel et al., A five week, polysomnographic assessment of zaleplon 10 mg for the treatment of primary insomnia. Sleep Med. **1**(1), 41–49 (2000)

C. Wang, G. Cunningham et al., Long-term testosterone gel (AndroGel) treatment maintains beneficial effects on sexual function and mood, lean and fat mass, and bone mineral density in hypogonadal men. J. Clin. Endocrinol. Metab. **89**(5), 2085–2098 (2004)

C. Wang, R.S. Swerdloff et al., Transdermal testosterone gel improves sexual function, mood, muscle strength, and body composition parameters in hypogonadal men. J. Clin. Endocrinol. Metab. **85**(8), 2839–2853 (2000)

G. Wang, H.M. Lee et al., Ghrelin – not just another stomach hormone. Regul. Pept. **105**(2), 75–81 (2002)

S.S. Wang, T.A. Good, Effect of culture in a rotating wall bioreactor on the physiology of differentiated neuron-like PC12 and SH-SY5Y cells. J. Cell Biochem. **83**(4), 574–584 (2001)

Y. Watanabe, H. Ohshima et al., Intravenous pamidronate prevents femoral bone loss and renal stone formation during 90-day bed rest. J. Bone Miner. Res. **19**(11), 1771–1778 (2004)

D.E. Watenpaugh, S.F. Vissing et al., Pharmacologic atrial natriuretic peptide reduces human leg capillary filtration. J. Cardiovasc. Pharmacol. **26**(3), 414–419 (1995)

W.W. Waters, S.H. Platts et al., Plasma volume restoration with salt tablets and water after bed rest prevents orthostatic hypotension and changes in supine hemodynamic and endocrine variables. Am. J. Physiol. Heart Circ. Physiol. **288**(2), H839–H847 (2005)

W.W. Waters, M.G. Ziegler et al., Postspaceflight orthostatic hypotension occurs mostly in women and is predicted by low vascular resistance. J. Appl. Physiol. **92**(2), 586–594 (2002)

D. Watt, L. Lefebvre, Vestibular suppression during space flight. J. Vestib. Res. **13**(4–6), 363–376 (2003)

W.B. Webb, Sleep as an adaptive response. Percept. Mot. Skills **38**(3), 1023–1027 (1974)

N.J. Wesensten, Effects of modafinil on cognitive performance and alertness during sleep deprivation. Curr. Pharm. Des. **12**(20), 2457–2471 (2006)

N.J. Wesensten, T.J. Balkin et al., Reversal of triazolam- and zolpidem-induced memory impairment by flumazenil. Psychopharmacology (Berl.) **121**(2), 242–249 (1995)

L. West, *Dream Psychology and the New Biology of Dreaming* (Charles C. Thomas, Springfield, IL, 1969)

M.H. Whitnall, C.E. Inal et al., In vivo radioprotection by 5-androstenediol: stimulation of the innate immune system. Radiat. Res. **156**(3), 283–293 (2001)

M.H. Whitnall, V. Villa et al., Molecular specificity of 5-androstenediol as a systemic radioprotectant in mice. Immunopharmacol. Immunotoxicol. **27**(1), 15–32 (2005)

P.A. Whitson, R.A. Pietrzyk et al., Effect of potassium citrate therapy on the risk of renal stone formation during spaceflight. J. Urol. **182**(5), 2490–2496 (2009)

P.A. Whitson, L. Putcha et al., Melatonin and cortisol assessment of circadian shifts in astronauts before flight. J. Pineal Res. **18**(3), 141–147 (1995)

M.L. Wiederhold, J.L. Harrison et al., A critical period for gravitational effects on otolith formation. J. Vestib. Res. **13**(4–6), 205–214 (2003)

J.R. Williamson, N.J. Vogler et al., Regional variations in the width of the basement membrane of muscle capillaries in man and giraffe. Am. J. Pathol. **63**(2), 359–370 (1971)

J.W. Wilson, S. Thibeault, et al., Issues in protection from galactic cosmic rays. Radiat. Environ. Biophys. **34**: 217–222 (1995)

J.W. Wilson, C.M. Ott et al., Space flight alters bacterial gene expression and virulence and reveals a role for global regulator Hfq. Proc. Natl. Acad. Sci. USA **104**(41), 16299–16304 (2007)

J.W. Wilson, C.M. Ott et al., Media ion composition controls regulatory and virulence response of Salmonella in spaceflight. PLoS One **3**(12), e3923 (2008)

J.W. Wilson, R. Ramamurthy et al., Microarray analysis identifies Salmonella genes belonging to the low-shear modeled microgravity regulon. Proc. Natl. Acad. Sci. USA **99**(21), 13807–13812 (2002)

S.M. Wimalawansa, S.J. Wimalawansa, Simulated weightlessness-induced attenuation of testosterone production may be responsible for bone loss. Endocrine **10**(3), 253–260 (1999)

C.D. Wood, A. Graybiel, Evaluation of sixteen anti-motion sickness drugs under controlled laboratory conditions. Aerosp. Med. **39**(12), 1341–1344 (1968)

C.D. Wood, J.E. Manno et al., The effect of antimotion sickness drugs on habituation to motion. Aviat. Space Environ. Med. **57**(6), 539–542 (1986)

C.D. Wood, J.E. Manno et al., Side effects of antimotion sickness drugs. Aviat. Space Environ. Med. **55**(2), 113–116 (1984)

C.D. Wood, J.J. Stewart et al., Therapeutic effects of antimotion sickness medications on the secondary symptoms of motion sickness. Aviat. Space Environ. Med. **61**(2), 157–161 (1990)

M.J. Wood, C.D. Wood et al., Nuclear medicine evaluation of motion sickness and medications on gastric emptying time. Aviat. Space Environ. Med. **58**(11), 1112–1114 (1987)

D. Woodard, G. Knox et al., Phenytoin as a countermeasure for motion sickness in NASA maritime operations. Aviat. Space Environ. Med. **64**(5), 363–366 (1993)

H. Wu, J. Huff, et al., Risk of Acute radiation Syndromes due to Solar Particle Events (NASA, Houston TX, HHC, 2008)

M. Xiao, M.H. Whitnall, Pharmacological countermeasures for the acute radiation syndrome. Curr. Mol. Pharmacol. **2**(1), 122–133 (2009)

J. Xiong, Y. Li et al., Effects of simulated microgravity on nitric oxide level in cardiac myocytes and its mechanism. Sci. China C. Life Sci. **46**(3), 302–309 (2003)

M. Yamaguchi, K. Ozaki et al., Simulated weightlessness and bone metabolism: impairment of glucose consumption in bone tissue. Res. Exp. Med. (Berl.) **191**(2), 105–111 (1991)

B.J. Yates, A.D. Miller, Physiological evidence that the vestibular system participates in autonomic and respiratory control. J. Vestib. Res. **8**(1), 17–25 (1998)

B.J. Yates, A.D. Miller et al., Physiological basis and pharmacology of motion sickness: an update. Brain Res. Bull. **47**(5), 395–406 (1998)

A.D. Yegorov, L.I. Kakurin et al., Effects of an 18-day flight on the human body. Life Sci. Space Res. **10**, 57–60 (1972)

J.J. Zachwieja, S.R. Smith et al., Testosterone administration preserves protein balance but not muscle strength during 28 days of bed rest. J. Clin. Endocrinol. Metab. **84**(1), 207–212 (1999)

G.K. Zammit, B. Corser et al., Sleep and residual sedation after administration of zaleplon, zolpidem, and placebo during experimental middle-of-the-night awakening. J. Clin. Sleep Med. **2**(4), 417–423 (2006)

J.E. Zerwekh, C.V. Odvina et al., Reduction of renal stone risk by potassium-magnesium citrate during 5 weeks of bed rest. J. Urol. **177**(6), 2179–2184 (2007)

J.G. Zhi, G. Levy, Kinetics of drug action in disease states. XXXVII. Effects of acute fluid overload and water deprivation on the hypnotic activity of phenobarbital and the neurotoxicity of theophylline in rats. J. Pharmacol. Exp. Ther. **251**(3), 827–832 (1989)

Y. Zhou, M.T. Mi, Genistein stimulates hematopoiesis and increases survival in irradiated mice. J. Radiat. Res. (Tokyo) **46**(4), 425–433 (2005)

S.R. Zwart, J.E. Davis-Street et al., Amino acid supplementation alters bone metabolism during simulated weightlessness. J. Appl. Physiol. **99**(1), 134–140 (2005)

S.R. Zwart, A.R. Hargens et al., The ratio of animal protein intake to potassium intake is a predictor of bone resorption in space flight analogues and in ambulatory subjects. Am. J. Clin. Nutr. **80**(4), 1058–1065 (2004)

S.R. Zwart, D. Pierson et al., Capacity of Omega-3 fatty acids or eicosapentaenoic acid to counteract weightlessness-induced bone loss by inhibiting NF-kappaB activation: from cells to bed rest to astronauts. J. Bone Miner. Res. **25**(5), 1049–1057 (2009)